THE FORCES

THAT DRIVE

SUBSURFACE

GEOLOGIC

PROCESSES

by Dr. T.C. Holmes

The Forces that Drive Subsurface Geologic Processes

by Dr. T.C. Holmes

First Edition

T.C. Holmes
Victoria, British Columbia, Canada

The sponsoring editor of this book is (TCL) Thomson Communications Limited. It was set in Perpetua, Times New Roman and Arial Narrow by TCL.

Canadian Cataloguing in Publication Data
Holmes, T.C. (Terence Charles), 1906-
    The forces that drive subsurface geologic processes

Includes bibliographical references.
ISBN 0-9683320-0-5

    1. Geodynamics.   I. Title

QE33.2.S9H64 1998   551.1   C98-900146-6

The Forces that Drive Subsurface Geologic Processes

TABLE OF CONTENTS

PART III: SOURCES OF HEAT THAT AFFECT THE EARTH /29

PART IV: CAUSES OF SUBSURFACE GEOLOGICAL PROCESS /55

PART V: CONCLUSION /91

ACKNOWLEDGEMENTS /95

REFERENCES /97

About The Author

Dr. Terence C. Holmes was born in Victoria, British Columbia, in 1906. He is the son of William C. Holmes who served as a judge between 1877 and 1902 in the Indian Civil Service. The Service comprised a select group of about 1,000 individuals who administered British rule in India for almost 100 years.

Dr. Holmes attended about three years of public and high school. He was then articled to the B.C. Land Surveyor in 1923 receiving his commission as B.C. Land Surveyor in 1927. From 1926 through 1927, he took a special course and also matriculated in 1927.

Dr. Holmes completed his Bachelor of Science in Mining Engineering at the University of British Columbia in 1932 and his PhD in Geology at the University of Chicago in 1936. He also studied for two quarters at the University of Minnesota, mainly under Dr. Frank Grout. During the next four years, Dr. Holmes gained a variety of employment experiences in geological work connected with mining, prospecting and diamond drilling.

In 1940, Dr. Holmes began work as a geologist at the Dome Mine, a gold mine in Ontario, becoming Chief Geologist in 1942. He stayed 31 years, drilling about 400 holes a year and driving 15,000 feet of horizontal tunnels and several thousand feet of raises for much of this time. During his last 15 years with Dome, Dr. Holmes was also Chief Mining Engineer and is credited with making key suggestions which led to important changes in mining methods.

It was several years after retirement in 1971 that Dr. Holmes began a more in-depth process of scrutinizing present geologic theories, and more specifically the forces that drive subsurface geologic processes. His research led him to a variety of geologic papers that referenced the low velocity zone. Over the next fifteen to twenty years, Dr. Holmes developed concepts and ideas around activity that occurred in the low velocity zone and how geologic processes were impacted. The result of his years of work experience and theorizing culminated in the development of this paper.

Dr. Holmes continues to theorize, speculate and write on *The Forces That Drive Subsurface Geologic Processes* from his home in Sidney, British Columbia.

Abstract

We know the age of the oldest fossils on earth. Projecting this length of time backward to when life first appeared and adding time for the accumulation of oceans in which life lived suggests the earth's surface has varied little in temperature for a long time. Two and one half billion years is suggested.

The theory of radioactive decay has to be changed to fit this. It is suggested that pressure reduces the rate of decay until, at a few tens of kilometres in depth, it is nil. The loss of radioactive material near surface by decay is made up by new material brought up from depth by magmas.

Heat is generated throughout the earth by internal friction caused by tides in the solid body of the earth. It moves outward. In the Low Velocity zone (LV), the heat causes a liquid fraction to form just above any contact.

The rocks of the earth tend to be arranged in layers with the denser rocks below. The denser rocks tend to be more basic and have a higher initial melting point. In the LV the rock heats until a liquid fraction or liquidus forms just above each contact. Its presence slows the seismic waves. When the liquidus becomes abundant enough in any horizon to be continuous, gravity pushes it up slopes in the horizon because it is lighter, and it forms a pool of magma at any apex.

A theoretical study of what happens then suggests the magma rises from the pool as narrow dikes. Its behaviour after that depends on its density and that of the surrounding rocks.

- Lighter magmas expand a short distance below surface forming batholiths and pushing the walls aside.

- Magmas of intermediate composition tend to erupt. *A reason is suggested why volcanoes tend to have periods of activity separated by periods of dormancy.*

- Denser magmas tend to spread out as sills at depth causing regional uplifts with little folding.

- The surface sinks above areas that material moves out of and rises above areas that material moves into.

Reason is given why the process in any one horizon in the LV increases in intensity, peaks, and then dies out. This accounts for geologic revolutions starting and stopping at certain places.

The activity in any horizon in the LV affects other horizons. Usually only one will be active in any area at any one time, except near the end of a cycle of activity.

It is suggested that the heat of fusion is increased by pressure and that this extra heat is given up as fused material moves into areas of lower pressure.

The huge variety of things that actually are seen on the earth's surface suggest that the picture that will be presented in Parts III and IV of this paper is only a broad and simplified picture of what goes on beneath the earth's surface.

Introduction

We understand the forces that drive geologic processes at and above the surface of the earth. The processes include air, water, ice movements and slight adjustments near surface in the solid. The driving forces are heat from the sun and gravity. Their effect is strikingly different in the atmosphere, the hydrosphere and the lithosphere.

In sharp contrast is our lack of understanding of the forces that drive subsurface geologic processes. Various solutions have been offered. However, the solution of problems is frequently more complicated than may appear at first glance. An example is the relative importance of the sun and moon in causing tides in the oceans. At first glance one might think their effect would vary directly as their mass and inversely as the square of the distance from the earth. Calculations on this basis show the sun's effect is about twice that of the moon. But, observation shows the opposite is approximately true. So man seeks an explanation. He comes up with one based on the fact that the earth subtends an angle of nearly 2° at the moon, but only about 0° 0′ 30″at the sun. *It seems plausible, but would it have been thought of at all had the problem not been identified? Is there possibly a more satisfactory solution that has not been investigated because thinking stopped as soon as something "plausible" was suggested?*

In how many cases has thinking stopped at an unsatisfactory answer as soon as something "plausible" was suggested?

TROUBLES WITH PLATE TECTONICS

In the theory of plate tectonics the earth's surface is considered to be divided into plates which move very slowly in various directions with respect to one another. In at least some versions of the theory, the plate is pictured as coming up from the depths of the earth on a slope of about 45^0, moving along the surface, and some thousands of kilometres away, going back down into the earth also on a slope of about 45^0.

I find this implausible for the following reasons:

1. *How did the plate form at depth initially? What was different about that material from that around it? What governs its thickness? ... its dip?*

2. *What force pushed the plate upward, sideways, downward?* Credence is sometimes given to the force of gravity. It is suggested that the rising part of the plate is less dense because it is hotter and the descending part is cooler because it has been on surface for so long.

   It seems doubtful that these changes in density would be significant if compared to the changes to be expected in rock density going deeper in the earth. At the very slow rate of movement indicated by the theory of plate tectonics, the plate and wall rock would have to be almost exactly the same temperature at any horizon due to conduction of heat. One might argue there was some source of heat in the rising plate: *But what is it?* Also this would imply a source of cooling in the descending plate!

3. If the force on the rising plate were pushing it upward and then pushing the horizontal plate along, one would expect a great deal of buckling just beyond the bend where the plate reached surface.

4. Solid rock does not easily slide over solid rock. Therefore, the proceeding point makes the process of plate tectonics doubtful.

5. Both the rising and descending plates are pictured as bending about 45^0 as they go around the corner at surface. An unlikely picture.

6. If gravity were the driving force, one would expect movement to be vertical and not on a 45^0 slope.

7. Consider the distribution of rocks in North America — the Canadian Shield, the Appalachian Mountains, the nearly flat sediments of the Central Plains, the Western Cordillera. *Could these have formed at depth in a plate and been moved up to surface and along to their present position by movement of the plate?*

8. Various writers show different ideas of what happens where they picture the plate going down into the earth. I have not noted one that allows for the rocks above the descending plate being raised by the plate pushing in below them.

9. The sides and bottom of the plates have to be thought of as fault surfaces with the plate sliding along them. *Why are these not the loci of many faults — especially the bottom?* I do not recall hearing of frequent earthquakes along these lines.

10. Plate tectonics does not answer many of the questions raised in Part I of this paper.

Some years after retirement, to help pass the time, I started to wonder about the problems with the theory of plate tectonics. I had no idea what the answer might be. So I did a good deal of random reading mostly in the University of Victoria Library. Based on what I learned and my practical experiences, a plausible hypothesis had gradually developed and is now quite advanced. Unfortunately increasing age prevents its further development. It is presented in its incomplete condition in this paper. Hopefully, others will take it up and improve it.

In putting this hypothesis into words, the ancient problem of *"which came first, the chicken or the egg?"* has plagued me. The reader's patience and indulgence is respectfully requested.

ABOUT THIS PAPER

This paper contains five parts.

PART I: BASIC QUESTIONS — outlines topics that must be explained, and poses a great many questions that are not answered by the theory of plate tectonics.

PART II: SCIENTIFIC BACKGROUND — summarizes a number of key scientific facts that have been well recognized for several decades. The hypothesis presented rests largely on these facts.

PART III: SOURCES OF HEAT THAT AFFECT THE EARTH — presents a concept on how heat may be generated in the earth.

PART IV: CAUSES OF SUBSURFACE GEOLOGICAL PROCESSES — reviews the actual behaviour of subsurface geologic processes and presents a concept of the causes of these behaviours.

PART V: CONCLUSION — summarizes the theoretical information presented.

PART I: Basic Questions

Unanswered Theoretical Questions

As stated earlier, I theorize that the hypothesis of plate tectonics has become popular for lack of a more plausible hypothesis. I note that plate tectonics fails to answer or even to address a number of important questions.

Nevertheless, before presenting another hypothesis, I present in point form a number of geologic events and occurrences that require explanation, and list questions that remain unanswered. If the point presented is followed by (✓), a plausible answer is suggested in Part III or Part IV of this paper. If the point is followed by ( ), a plausible answer has not yet been developed.

■ **In the central plains of North America,** a large area of flat, undeformed, shallow water sediments exists, suggesting widespread settling of the whole area with only negligible tilting over a very long time during sedimentation. Interpretation of the stratigraphic evidence shows that there has been a lot of raising and lowering of various parts of the area, at various times during sedimentation, but there has been negligible folding — just slight tilting.

Often the volume of material that moved in or out was measured in tens or hundreds of thousands of cubic kilometres and the distance moved must have been over one thousand kilometres.

What moved in below to cause the uplifts and out to permit the settlings? (✓) At what horizons did these movements occur? (✓) Was the material liquid or solid when moving? (✓) What forces caused all these movements? (✓) They must have been extremely large.

■ **These sediments extend westward into an area of geosynclinical sediments,** sometimes well over ten thousand metres thick, which were deposited in a relatively short time. Subsequently, this area became the site of uplift, and much folding, and faulting. Igneous activity was common in the Cordillera immediately to the west. Again, a volume of roughly the same order must have moved.

What left from below the area where the geosyncline was deposited to make room for it to settle into? ( ) Where did that material go? ( ) What moved to make room for it? ( ) Why the change to much thicker sediments? ( ) What forces caused all these large movements? ( ) How is it related to the formation of the rest of the Cordillera to the west? ( ) ... to the Rocky Mountain Trench? ( ) Why did settling change to uplift? ( ) Why did this become the site for igneous activity? ( )

■ **Igneous rocks outcrop over considerable areas of the earth's land surface.** They are believed to be the result of the cooling of magmas which rose from great depth.

Where did these magmas come from? (✓) Why did melting occur where it did? (✓) Why did it stop? (✓) As a rule, the movements of magma seem to have been intermittent rather than steady: Why is this? (✓) What moved into the space the magma left? (✓)... and into the space that material came from? ( ) What force caused all these movements? (✓)

■ **Some igneous rocks are intrusive; others are extrusive.** Almost all intrusives are in batholiths and other large bodies. The plutonic rocks seen in areas of Palaeozoic and younger rocks occur in mountainous areas.

■ **Most intrusives are near or in geosynclines which have been compressed, raised, folded, and faulted.** They are almost entirely acidic. Only a negligible proportion are basic.

■ **In the Precambrian Shield of North America, acidic batholiths are much more abundant than in areas of younger mountains.** They are thought to be the roots of ancient mountain ranges, now reduced by erosion to peneplains.

Why are geosynclines, mountain building and batholiths associated? ( ) Why are batholithic rocks much more abundant in Precambrian areas than in more recent mountainous areas? (✓) What caused intrusion to start? (✓) ... to stop? (✓) What governs the size of intrusions? (✓) What happened to the rocks that were where the batholiths now are? (✓) What replaced the spaces at depth that the magmas left? (✓) Why are batholiths dominantly acidic? (✓) What forces caused these movements? (✓) What started the forces working, and why did they stop?(✓)

■ **The volumes and distances involved in the formation of the batholiths of the Precambrian Shield of North America are very roughly of the same order as the volumes and distances noted above re the development of the Western Plains and their westward extension.**

■ **Most extrusives emerge through fissures and form extensive deposits of the plateau lava type.** Many of these are recent submarine flows. A small proportion rise through pipes and form volcanoes.

The author recalls lectures in the graduate school at the University of Minnesota by F.F. Grout, over 60 years ago, at which time he said about two thirds of extrusives were basic, and only one third acidic. With today's knowledge of the large volumes of basic lavas extruded beneath the oceans, it seems that "two thirds" may be considerably "too low".

Why are extrusives mainly basic? (✓) *Why the difference between the proportions of acid and basic rocks on land and under water?* (✓) *Why do some rocks form extrusives rather than intrusives?* (✓) *What forces cause their movement?* (✓)

■ **Eruptions in individual volcanoes usually follow a pattern.** A series of eruptions occur closely spaced in time. Then the volcano is dormant for a long time, possibly hundreds of years, followed by repetitions of periods of eruption and dormancy, to finally become dormant for good. Earthquakes tend to follow a similar cycle.

What is the cause of this pattern? (✓) *Do plateau lavas follow a similar pattern?* ( ) *Do intrusions follow a similar pattern?* ( )

■ **A number of large sills are known near the earth's surface.** Their existence is not realized readily from the surface.

Are there many more large sills, possibly much larger, that are so deep that they are never seen? Why do some magmas consolidate as sills? (✓) *Does their intrusion account for the sediments being above sea level?* (✓)

- **Schists are quite common in some areas, especially in the Precambrian.** The schistosity is usually steep. *What caused it?* (✓) *Why is it usually steep?* (✓) *What are the conditions necessary for the formation of schists?* (✓)

- **The idea that magma would form as a result of melting of the entire rock in one block is implausible.** Data accumulated by Bowen on cooling of molten silicate mixtures indicates that there is probably a difference of some two hundred degrees between the temperature of first crystallization and final crystallization. It seems reasonable to think that if a magma was formed by melting rock — in place — until the central part of the rock was completely molten, the surrounding material would grade outward in temperature, and proportions of molten material, until 100% solid was reached.

  Since the temperature change would be somewhere around 200° the gradational zone would be large — at least some miles — and the magma would be less dense than the surrounding rock. This seems unlikely for three reasons.

  (i)   I know of no reason why heat would concentrate in this small block.

  (ii) When rock melts it expands so the magma would be less dense than its surroundings. According to the principle illustrated in Figure 1A *(see page 18)*, the pressure differential between the liquid and the surrounding solid would be great enough to cause the liquid to start rising well before complete melting was reached.

(iii) Igneous action is associated with geologic revolutions which start and stop.

What explains why melting would start at one place and later stop? (✓)

■ **The solid crust of the earth is quite thick, contrary to earlier belief that the solid crust might be quite thin.** In 1914, Michaelson ran a long experiment which enabled him to calculate that the solid crust is quite thick. The results of Michaelson's experiments were confirmed by later studies in seismology.

■ **The solid earth has tidal effects which correspond in frequency and magnitude with the tides in the ocean, but which are considerably out of step with the timing Michaelson expected.** Michaelson discovered this while conducting experiments to determine the thickness of the earth's solid crust, but he could think of no plausible explanation for this lag.

What is the explanation? (A possibility, albeit requiring further development, is suggested later in this paper.)

■ **The rate of radioactive decay is thought to be unaffected by pressure.** If this is so, the rate of heat supply from this source is declining. But the length of time life has been on earth indicates that surface temperatures have been very stable for a long time.

What is the explanation of this apparent contradiction? (✓)

PART II: Scientific Background

Part II reviews several scientific ideas which are well established and which are important to understanding the hypothesis presented in Parts III and IV.

Throughout this paper, I will use the terms "liquidus" and "solidus" for the two intimately mixed phases of partially melted rock. If the liquid phase is so abundant that the solid phase is not continuous, the whole is a fluid and is classed as "magma".

DIFFERENTIAL PRESSURE WITH DEPTH

Information in Part II is pertinent to the ideas and hypothesis presented later in this paper. This section explains a principle that is very important in understanding the forces that work at depth in the earth, but which does not seem to be understood by some.

The principle applies to differential pressure at depth. It explains the forces that cause mountain building and related geologic phenomena such as the movement of magmas, the folding and faulting of sedimentary mountains like the Rockies, the uplift of the sediments in the Western Plains and other major movements in the earth. It explains what causes salt domes and possibly pingos.

Differences in density in the earth are quite small. Much larger differences are used here in order to be obvious in these small figures.

Differential pressure between a solid and a contained liquid varies with depth. The application of this principle, as it relates to the earth, is illustrated in Figure 1A.

Figures 1A — 1D: How the differential pressure between a solid and a contained liquid varies with depth.

An artificial situation designed to illustrate an important principle. Pressure in the liquid equals that in the solid at the level of the membrane.

The series of figures that follow show how pressure on the wall inside and outside the tank vary with depth as liquids and granules of different densities are used.

The relative densities between the different elements is greatly enlarged over those found in nature. This is done to illustrate the principle.

If natural densities were used, the graphs would scarcely diverge. The differences only become significant when applied over the great heights that are found in nature.

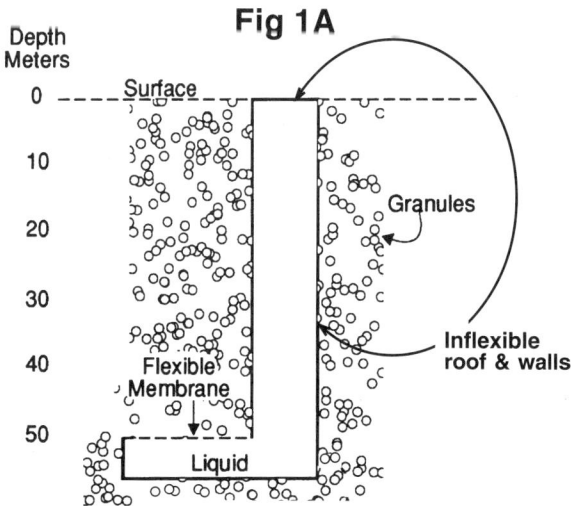

Fig 1A

Figure 1A shows a totally enclosed tank filled with liquid. Its walls and roof are solid except for a horizontal, flexible membrane at a depth of 50 metres. The surrounding area is filled with granular material with no tensile or shear strength. At a depth of 50 metres the granular material exerts pressure on the membrane. This pressure is transferred to the liquid, which is therefore under the same pressure at this depth.

Pressure in the solid at any point above 50 metres depends on the depth and the density of the granules. Pressure in the liquid at any point above this depth is reduced from the pressure at the membrane by the pressure exerted by a column of liquid of that height. The important point to note is: If the density of the liquid is less than that of the solid, the resultant pressure is outward on the walls of the tank.

Figures 1B, 1C and 1D graph pressure in both solid and liquid to show how their relation varies with depth as materials of different density are used.

In Figure 1B, the liquid is density 1.0 and the solid is density 2.0. The pressure outward on the walls of the tank, at any elevation, is measured by the horizontal distance between the two lines. It is always outward and increases steadily going up.

In Figure 1C, density of the granules above 30 metres is reduced to 1.0. The changes this causes are:

1. Pressure in both granules and liquid is reduced at every level.

2. Differential pressure outward increases in the same way up to the 30 metre level but from there to the top it remains unchanged.

Fig 1B

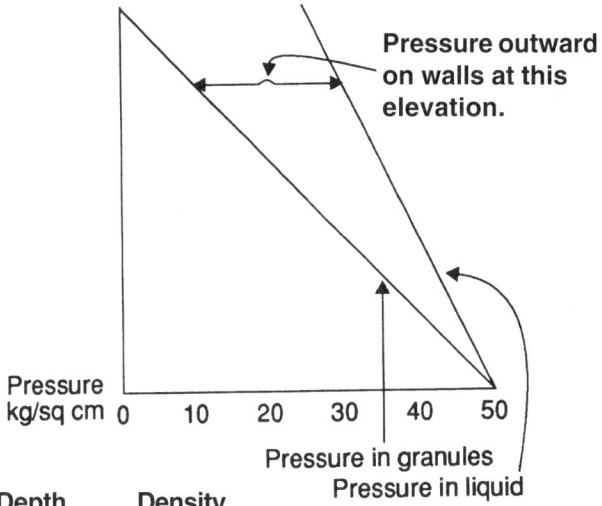

Pressure outward on walls at this elevation.

Pressure kg/sq cm

0   10   20   30   40   50

Pressure in granules
Pressure in liquid

	Depth	Density
Granules	0-50	= 2.0
Liquid	0-50	= 1.0

Fig 1C

Pressure outward on walls at this elevation.

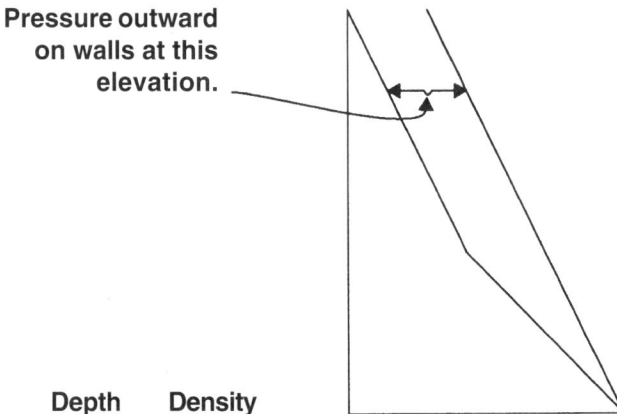

	Depth	Density
Granules	0-30	= 1.0
Granules	30-50	= 2.0
Liquid	0-50	= 1.0

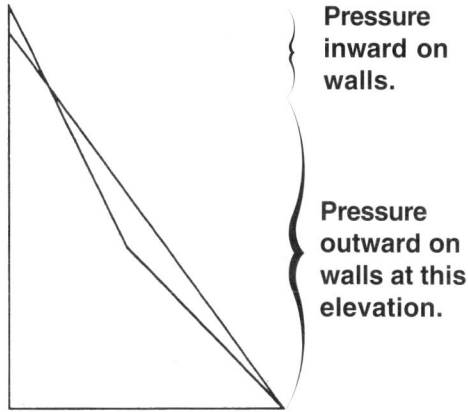

Fig 1D

	Depth	Density
Granules	0-30 =	1.0
Granules	30-50 =	2.0
Liquid	0-50 =	1.5

In Figure 1D, in addition to the change shown in Figure 1C, the density of the liquid is increased from 1.0 to 1.5 which is between that of the densities of the two types of granules used.

The additional changes this causes are:

1. The outward pressure increases at a much slower rate going upward to the height where the density of the granules change.

2. From here upward the density of the granules is less than that of the liquid so the outward pressure decreases going upward until it is NIL. From here up the pressure is inward.

Thermal Gradient

Thermal gradient is the rate at which temperature in the earth increases with depth. It has been measured directly down to depths of 5 km or so at numerous places in the earth.

At each place where thermal gradient has been measured, the rock temperature increased with depth at a rate which has been assumed to be quite uniform. *(A more detailed analysis of thermal gradient is included in Part III: Sources of Heat that Affect the Earth.)*

This analysis indicates that normally the thermal gradient has decreased greatly before the LV is reached, otherwise rock would be molten. The observed thermal gradient varies considerably from place to place. It is low in the old Precambrian Shield areas and tends to be high in areas of recent tectonic activity.

The depth of the zone where we can make direct observations is only a small fraction of the depth to where melting might occur and is usually much less than one thousandth of the depth to the earth's centre. So estimates of the depth where molten material may be expected, based on projections of the observed thermal gradient, are of little value.

What Happens When Rock Is Heated?

A great deal of careful and well organised laboratory experimentation was done at the Carnegie Geophysical Institute under the directorship of N.L. Bowen. Experiments to show what happens as magmas cool and crystallize started in the 1920s.

At that time the order of crystallization was a matter of major interest to petrologists. Various mixtures of two or three basic chemicals were melted and allowed to cool. The course of cooling was monitored and the products were examined in detail. From this information the temperature of formation of the solid products, the ranges of their stability, the changes of their composition with changing temperature, and other phenomenon were observed.

Even in these mixtures, which were relatively simple compared to natural magmas, Bowen found the results were extremely complex. They showed that in general:

1. The first crystals formed were quite different in composition from the average of the melt.

2. When reacting with the melt, the composition of crystals usually changed if cooling progressed slowly.

3. The proportion and composition of almost all minerals varied as cooling proceeded.

4. Some of the first-formed minerals were totally resorbed or changed to other minerals.

5. The difference in temperature between first crystallization and complete crystallization was large.

Based on the study of these results and of the textures found in igneous rocks and other field observations, Bowen developed the concept of a general order of crystallization of the minerals in igneous rocks.

This concept is known as *Bowen's Reaction Series* and is familiar to all petrologists. Crystallization following this general order results in more common igneous rock.

Bowen also suggested processes that might take place under special conditions, to be expected in the earth, under which cooling would result in separation of certain phases, and thus account for some of the unusual rocks that are found.

Some writers have speculated in general terms on large effects that mineralizers might have in the sequence of crystallization. Bowen's analysis of this problem led him to conclude that it is improbable that mineralizers would be present in sufficient quantity to have more than a slight effect, except locally where special conditions may exist.

The most important results, in connection with the hypothesis developed in this paper, are the inferences that:

1. Complete melting would only take place over a considerable range of temperature.

2. After the initial melting temperature (IMP) is reached, liquid and solid phases coexist, which under suitable conditions might separate, yielding products of a different composition.

EFFECT OF PRESSURE ON MELTING POINT OF ROCKS, ETC.

According to Grout *(Petrology and Petrography, McGraw-Hill, p. 155)* some experiments have indicated that the melting point of rock increases about 2.5 to 5°C for each kilometre of depth in the earth but that results are uncertain. Grout also says that general principles suggest that the effect per kilometre of depth becomes less going deeper.

In view of the variety of minerals and rocks, wide local variations occur. At a 100 km depth melting temperatures may be increased a few hundred degrees due to this factor. The increase between a 100 and 200 km depth is probably less. At present, what the actual figures are is guesswork. Precise figures are not essential to the development of the hypothesis outlined in Part IV of this paper.

DENSITIES OF ROCKS & MAGMAS

Rock increases in volume when melted, with a corresponding decrease in density. In his publication, Petrology and Petrography *(p. 193)*, F.F. Grout says the amount of change has been estimated at about 6% for basic rocks and 9% for acidic rocks, but these figures are uncertain because the effect of high pressures on these percentages is not known.

In what follows, it is assumed there is a decrease in density of this order on melting at depth. The exact percentage of decrease is not essential to the hypothesis developed in this paper, but the relative density of rocks and magmas is important. Typical examples of density of solid rocks and their liquid states under surface conditions are noted in Table A. At depth the densities would increase, but it is assumed that a change in density would not greatly affect the relative values.

Table A: Typical densities of solid rock and its liquid form.

	Density of Solid	Increase in volume by Melting	Density of Liquid
Granite	2.66	9%	2.42
Diorite	2.87	7% assumed	2.67
Gabbro	3.00	6%	2.82
Peridotite	3.23	5% assumed	3.07
Shale	2.69		
Limestone	2.76		

The significance of the numbers in Table A is not their precise value, but how the densities of the liquids compare with those of solid rocks of the same, and different, compositions. When applied over the great heights envisioned in Part IV, the differences in total pressure between rocks and magmas becomes significant.

HEAT FACTORS

The specific heat of silicates is about 0.2 to 0.3 calories per gram; their heat of fusion about 80 to 110 calories per gram at atmospheric pressure. Therefore, if a silicate is being fused at atmospheric pressure, about a quarter to a half as much heat has to be added to fuse it as was required to heat it from surface temperature to the fusion point. Presumably the opposite applies on cooling. *(These figures may be substantially different at the high pressures where molten material is formed).*

EFFECTS OF CHANGES IN PRESSURE

Movement of material at the surface by erosion or sedimentation affects the height of the surface and the pressure in the areas beneath it. The depth of cover affects the temperature beneath it, and thus has side effects.

The effect of any change of pressure at the surface spreads out downward. Similarly, the effect of changes of pressure at depth spreads upward. So either effect soon becomes almost negligible unless the volume that moved is large.

Any change of pressure or temperature at depth affects the volume of rock there and is reflected all the way to the surface. Also, any movement of either liquids or solids at depth is reflected at the surface.

If the rock were free to expand sideways, expansion would occur equally in three directions. But the solid rock around any expanding block severely restricts any expansion sideways. So practically all expansion must be upward. Therefore, in this situation, the index of vertical expansion is about three times the index of volumetric expansion.

MOVEMENT IN LIQUID VS. SOLID STATE

Both liquids and solids have approximately fixed volumes, but changes in shape are produced in both when pressure and temperature change. Liquids change shape readily, whereas solids resist forces tending to change their shape — often very strongly.

Also, when the force is removed, liquids have little tendency to resume their original shape, whereas solids usually resume it promptly, provided the elastic limit was not exceeded.

The result is that large volumes of liquids will move readily through narrow and contorted channels, whereas solids will not.

THE LOW VELOCITY ZONE (LV)

Seismography indicates that the earth is solid down to about 2900 km. Earthquake waves passing through this part of the earth move faster with depth everywhere except between about 100 and 400 km below surface. Thickness and depth appear variable. In this zone the change in the wave velocity varies irregularly with depth. At most places it decreases, hence the name. Beneath the Canadian Shield it exhibits little change with depth. Reason for its existence does not appear to have been offered.

The hypothesis, and most significantly important part of this paper, deals with the make-up of the Low Velocity Zone, what happens in it, the forces produced there, the results caused, the different effects on rocks of different composition and density, and the changes that occur over time. *The hypothesis can be found in Part IV, starting on page 55.*

PART III: Sources of Heat that Affect the Earth

SUMMARY OUTLINE

Historical geology shows there has been intermittent igneous action on the earth as far back as geological records are capable of revealing. *Is this indicative of a large proportion of the earth's life to date?* This record makes it seem questionable whether the frequency of igneous action is increasing or decreasing over the long term.

Part III suggests the possible sources of heat responsible for the continuing generation of magmas.

THE SUN

The only obvious source of heat is the sun, and of course this source of heat is not within the earth. The sun directly affects all the earth's surface. It heats the area directly below it more than areas where it strikes the earth obliquely. As a result of the daily and annual movements of the earth, the area being heated by the sun is constantly changing and results in uneven heating which:

■ drives the winds of the atmosphere,

■ provides energy for the evaporation of water (which is returned when condensation occurs),

- provides the energy for growth of plants and, indirectly, animals,

- affects water temperature (causing changes of density which are a cause of currents), and

- produces many other effects near the surface.

The sun's direct heating effect penetrates the solid earth for only a few metres. It does not penetrate to depth. The sun's influence on conditions at depth is confined to a blanketing effect. Below this "blanket area," the earth gets warmer with depth from sources of heat within the earth.

The remainder of Part III is dedicated to discussing the sources of heat within the earth.

HEAT FROM RADIOACTIVE DECAY

Less than a century ago physicists thought the earth must be cooling, since the sun does not supply enough heat to maintain the earth's temperature. But palaeotological evidence indicated temperatures on earth had changed little for far longer than physicists thought possible. Consequently, geologists puzzled: *Where did the heat to maintain the earth's temperature come from?*

Later, radioactivity was discovered. Radioactivity does not emit enough heat to be noticeable in hand specimens, but is so widespread that the low heat conductivity of rock combined with the blanketing effect of deep cover makes it very significant at depth. Indeed, the question to answer became: *Why is the earth not becoming hotter?*

This was particularly puzzling since it was believed that pressure and temperature did not affect the rate of decay. To answer this question, it was assumed that radioactive material is concentrated in the upper layers, and that it is much sparser or absent at depth.

The assumption noted above answered the question for present-day conditions. But if conditions back in time are considered, difficulties in reasoning appear *(see Figure 2)*.

For example, if the quantity of radioactive heat generated by today's main contributors was doubled approximately every billion years, the rate of heat generation from today's main producers within the earth would have been much higher in the past.

Figure 2: Variation of heat from Radioactive Decay.

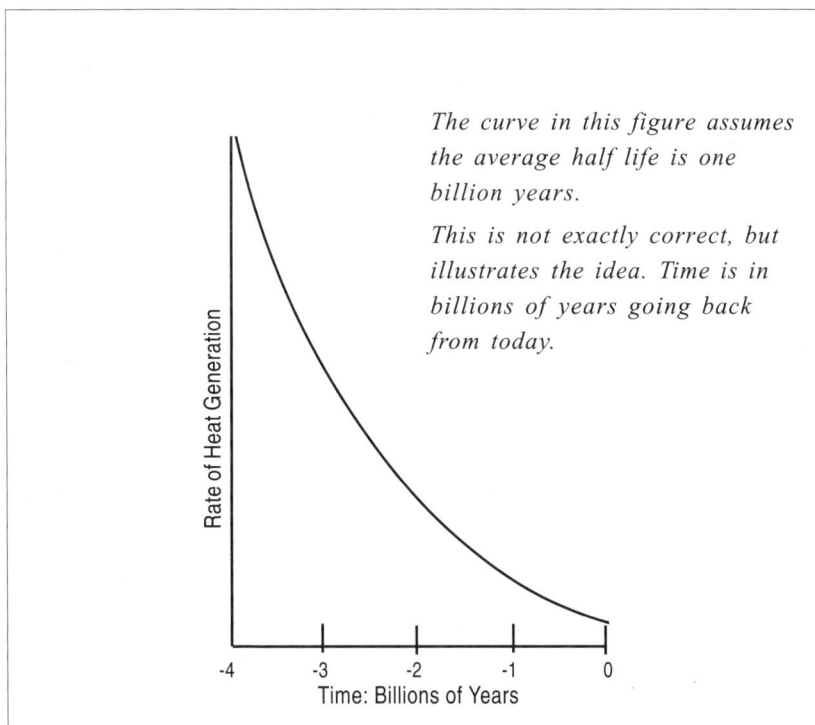

The curve in this figure assumes the average half life is one billion years.

This is not exactly correct, but illustrates the idea. Time is in billions of years going back from today.

Rate of Heat Generation

-4   -3   -2   -1   0
Time: Billions of Years

It would have made the surface so warm that oceans would have been too warm to permit life to exist in the distant past. If one considers that some radioactive elements have a considerably shorter half life, the discrepancy between present and past temperatures must necessarily be even greater. An alternative hypothesis is offered on the following pages.

A satisfactory hypothesis has to explain the facts that appear to be true. Bending the facts to fit the hypothesis is unsatisfactory, but may be difficult to avoid. A point that is suggested to be true concerns the age of the oceans. Various conclusions follow from this truth.

Radioactive dating indicates that the Paleozoic period started over half a billion years ago. At that time, animal life had developed hard calcareous parts. Before this period, life must have evolved through a great many stages starting with:

- **the change from inorganic material to the first molecule of organic life** — a huge step which probably took numerous attempts and a very long time to establish.

   The first molecule was a form of elementary plant life which took in carbon dioxide and gave off oxygen. Over time this life-form became more abundant, leading to the development of more complex life-forms.

   Initially, reproduction was by cell division. Later, another huge step took place when organisms with two sexes appeared.

- **the appearance of elementary animal life**. This was another huge leap towards the evolution of more complex life-forms. Since animals take in oxygen and give off carbon dioxide (the opposite of plants), presumably elementary animal life did not appear until plants had given off enough oxygen (i.e., making up a significant proportion of the atmosphere) to support animals.

This process required an abundance of plants and time. With still more time, animal life became even more complex and finally developed calcareous (hard) parts. Our knowledge of this process is supported through animal remains preserved in the fossils of the Cambrian period. Also, all limestones are now believed to be accumulations of the hard parts of animals, so these appeared long before the Cambrian.

Any of the development described above would have taken place in the ocean. Thus, the temperature at the earth's surface must have permitted the existence of oceans during all this time. Furthermore, prior to the appearance of the first life on earth, oceans had to accumulate. This process must have taken a great deal of time. Also, the temperature of the oceans probably varied little from what it is today, as it is unlikely that ocean life would survive any great temperature change.

Therefore, we can conclude that the temperature at the earth's surface has varied little for a very long time, possibly two or three billion years or more. It follows that the rate of heat supplied to the earth has also remained fairly consistent. This does not fit with present opinions or ideas about radioactivity.

An alternative hypothesis (presented as follows) regarding radioactivity and the rate of decay, does not conflict with the facts discussed on the previous pages.

The rate at which radioactive material decays is reduced greatly by conditions encountered above the LV. The rate of decay decreases going deeper which results in the thermal gradient decreasing enough that the temperature at depth varies somewhat as indicated in Figure 3, on the following page. The rate of decay must have remained about the same during and well back beyond the existence of life, while oceans were accumulating. This seems to require that the supply of heat from radioactive decay stayed pretty steady all this time.

Figure 3 suggests that the decay of radioactive material must be confined to the upper few tens of kilometres of the crust, otherwise temperature at the LV would be so far above the initial melting point that there would be a lot of molten material there. Since "S" waves pass through this zone, it seems this is not so. The following explanation is offered.

Conditions in the earth, below a few tens of kilometres, prevent radioactive decay. Decay of radioactive material near the surface takes place steadily over time, but there is no significant decrease of radioactive material near the surface because the loss has been made up by radioactive material in igneous rocks. This material was brought up in magmas from the area deep in the earth where they formed. This area must have been so deep that the rate of radioactive decay there was nil or certainly negligible. The magmas moved it from where conditions prevented decay to where the conditions permitted decay. This kept up the proportion of radioactive material in the horizons where conditions permit decay. However, presumably the supply at great depth will eventually run out.

Figure 3: How Initial Melting Point (IMP) varies with depth.

A generalized picture of how IMP and temperature vary with depth.

Actual position of all these curves and the height of the steps varies somewhat with both time and place.

Estimated temperature and thermal gradients at depth in the earth to the bottom of the Low Velocity Zone (LV), below two places, one with high initial thermal gradient near the surface, the other where it is quite low near the surface.

Figure 3 suggests:

1. *how initial melting point (IMP) varies with depth in the earth.* *The various steps indicate where rock composition changes in the LV.*

2. *how temperature varies with depth in the earth.* *The slope of this line is a measure of the thermal gradient. The curves move up or down with changes of temperature.*

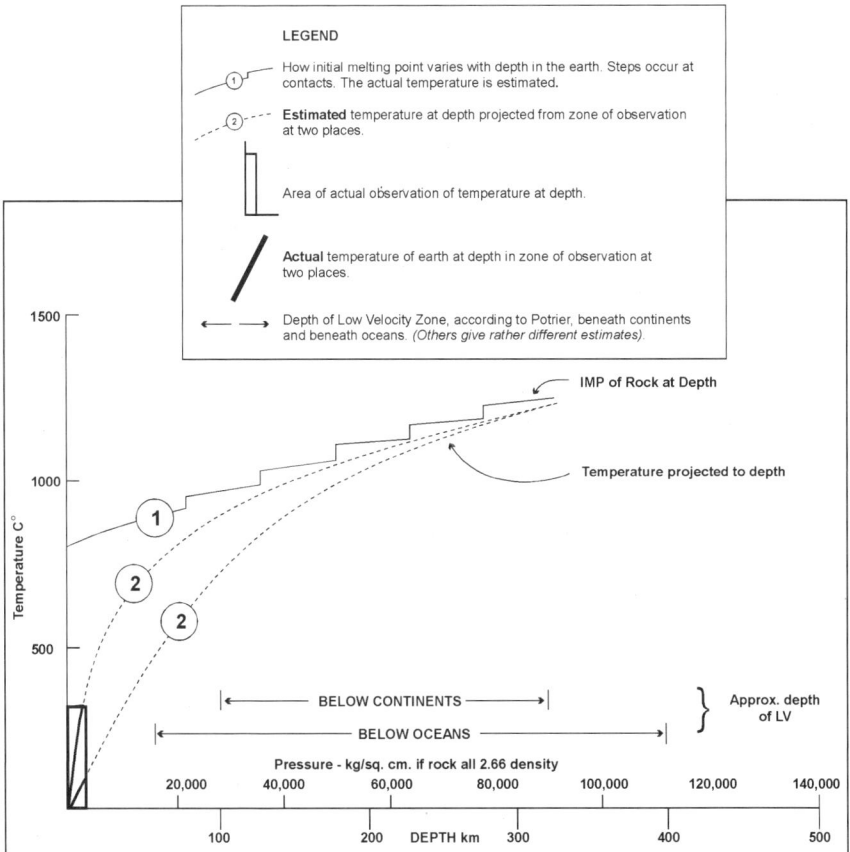

THE FORCES THAT DRIVE SUBSURFACE GEOLOGIC PROCESSES **35**

Heat by Internal Friction

It is suggested that another source of heat exists which produces heat in a much wider spread area. This heat source is internal friction which very slowly generates heat throughout the earth. Internal friction results from the body tides caused by the gravitational attraction of the sun and moon on the earth as it rotates on its axis. These were reported by Michaelson in 1914.

The different effects of internal friction on the solid outer shell, the liquid outer core and the solid inner core are discussed in the next section.

Gravitational Attraction of the Sun and Moon

Conditions in the earth's deep interior are inferred from studying seismic waves, what is seen on the surface of the character of rocks and their specific gravity, information from artificial satellites, and what the total weight and size of the earth is known to be. This evidence indicates that the earth is divided into various zones, which are described by Peter Wyllie in his book titled: *The Way the Earth Works* (1976).

These zones are:

1. The "shell" which is almost all solid. It extends down to about 2950 km and includes some subdivisions.

2. The "outer core", which is liquid. It is inside the shell and extends down to about 1400 km from the earth's centre.

3. The "inner core", where the earth is again solid.

When considering the effect of the gravitational attraction of the sun and moon on the earth, it is appropriate to divide the earth into three main zones. Seismic evidence indicates a change of state at each zone boundary. There is also a large increase in density going downward at each zone boundary.

The sun and moon both exert a gravitational attraction on the earth which acts on every particle in the earth in proportion to its density. At the ecliptic the pull of the moon is about 7% greater on the near side of the earth than on the far side. The pull of the sun is only about .02% greater. They differ because the sun is so much further away from the earth than the moon. The direction of the pull changes through 360° daily, due to the earth's rotation.

The amount of the attraction on each particle also changes slightly, due to the changing distance to the attracting bodies over time. Substantial changes in the net amount of the attraction are caused by the monthly changes in the relative positions of sun, moon and earth. At new moon they pull almost in the same direction, at full moon in almost opposite directions, and in between, the direction and magnitude of the pull is the resultant of the two and is smaller than at new moon. Also the distance to both the sun and the moon vary over time. These factors result in the intensity of their pull constantly changing a little.

Regarding the effect of generating heat within the earth, the significant difference between the force exerted on every particle in the earth by the pull of the sun and moon and the far larger pull of the earth's gravity is illustrated in Figure 4 *(see page 39)*.

The ovals in Figure 4, on the following page, represent cross-sections, at three hour intervals for 24 hours, through a particle that would be spherical if it were not for the pull of sun and moon. This pull distorts the particle into an ovoid shape whose long axis always lies in the same direction. The material of the particle rotates along with the earth. The straight line through each oval represents a line in the particle, which rotates with the earth and whose angle to the long axis of the oval is constantly changing. Also, its length is constantly changing.

If it were not for the pull of the sun and the moon, the particle would not change shape during the day and the length of every line through it would not change. For the purpose of illustrating the idea, the oval shape is hugely exaggerated. Actually, the difference from a sphere is so small that when Michaelson measured the body tides in the earth, quite an elaborate and refined setup was needed.

In contrast, the pull of the earth's gravity on every particle is always towards the earth's center so it has no tendency to change the shape of the particle as the earth rotates.

The pull of sun and moon results in tides throughout the earth. Heat is generated throughout the earth by the internal friction that results from these tidal effects. What happens in each of the three zones differs appreciably, and is described separately in the three following subsections.

Figure 4: How heat is generated all through the solid parts of the earth by internal friction.

This figure is a cross-section through the earth's center in the plane of the ecliptic.

At new moon, the pull of the sun and moon deform the earth so that every block in it that would be spherical without this pull is deformed into a prolate spheroid whose long axis nearly points toward the sun and moon.

This figure shows the same spheroid at 3 hour intervals over a 24 hour period. The constantly changing shape generates heat by internal friction.

An imaginary line is added through the spheroid to emphasize what happens. It rotates with the spheroid, so its length is constantly changing.

The shape of the spheroid takes a certain amount of time to adjust to the new direction of pull as it rotates. The direction of this lag is shown, but its size is merely guessed.

The amount the spheroid is distorted is vastly exaggerated. The actual amount is minute.

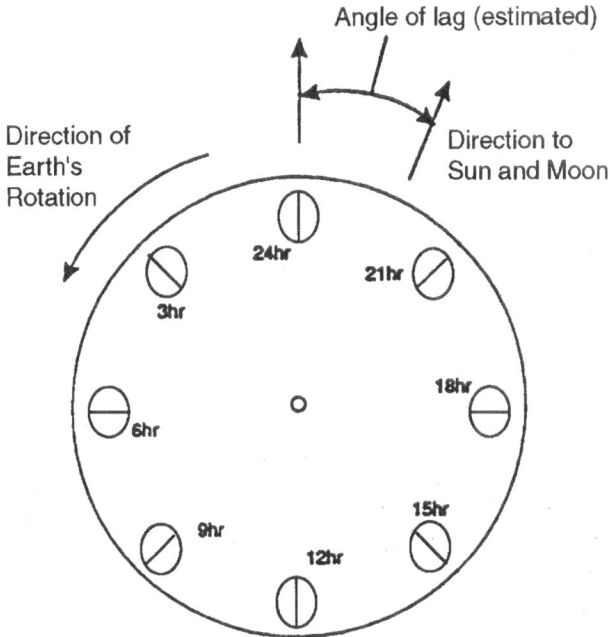

Effects in the Solid Shell

The attraction of both the sun and the moon act throughout the shell. At new moon they pull in nearly the same direction, so every block in the shell which would normally be spherical without that influence, is instead deformed into an approximate ovoid. The long axis of the ovoid points roughly toward the sun and moon at all times, so it stays in nearly the same direction. Since the pull of the moon in particular is greater in that part of the earth which is closer to the moon, the degree of deformation varies with depth. Therefore, the word "ovoid" is used instead of "oval".

The material of the ovoid is rotated about 360° every 24 hours because it is part of the rotating earth. Consequently, the material of the ovoid is continually flexed in the manner indicated in Figure 4. In this figure the amount the ovoid varies from a circle is vastly exaggerated in order to illustrate the concept. In fact, the variation is minute. This can be judged by the sophistication of the experiment Michaelson set up to measure it

It appears that the long axis of the ovoid should lag behind the direction to the sun and moon, since it takes time for the distortion to fully change in response to the force applied. The lag shown in Figure 4 is in the right direction, but its size is merely guessed.

1. The distortion decreases with depth as pressure increases.

2. If the same amount of work is done to distort rock at every depth, then the amount of heat generated may be equal at every depth even though the degree of distortion is different.

3. *Is the amount of work done independent of depth?*

If the amount of work gets less with depth, the amount of heat generated will decrease with depth. This means the rate at which heat is generated will decrease with depth. If this is so, conditions likely to generate liquidus will become more probable as the surface is approached; this helps to explain why the LV is located where it is.

4. *What affects the degree of distortion of the ovoid?*

   (i)   amount of pull of the sun and moon

   (ii) rigidity of material which is affected by:

       (a) temperature

       (b) pressure

       (c) type of material at that spot

Every particle in the shell is distorted in the same manner at any moment, so the total shell must be distorted to approximately the same ovoid shape, causing the surface to tilt slightly at some places. At the poles and equatorial plane of the ovoid, the surface is parallel to the spherical shape the earth would have if these pulls were absent. So, tilt there is zero. The tilt is at a maximum on the circles that go round the ovoid about halfway between its poles and equator. The axis of the ovoid is perpendicular to the plane of the earth's orbit so its poles and equator are some distance from those of the earth.

As noted earlier, Figure 4 shows the changes in the shape of the ovoid over 24 hours, at new moon. The earth is continuously being flexed, but the pattern varies through the month. If any object is flexed or bent, some heat is generated by internal friction. This flexing is generating heat continuously throughout the shell. The shell is more rigid at depth, so distortion is less. But since the force exerted at depth is the same, it is suggested that the rate of heat generation there may also be the same. Since the resistance to distortion is higher, more work has to be done per unit of distortion. *Is this suggestion correct?*

Effects in the Outer Core

Seismic evidence indicates that the earth is liquid from about a 2950 to 5000 km depth. It is much denser than the shell. This zone is known as the outer core. This implies a large change in chemical composition and considerably lower range of melting.

Gravitational attraction of the sun and moon act on every particle in the liquid outer core in a direction that varies 360° daily as the earth turns. This causes waves of very slight additional outward pressure to pass throughout the outer core daily. Also each wave in the outer core causes a wave of slight additional compression in the adjacent layers of the shell followed by relaxation as the wave passes on. The earth turns so rapidly that this effect probably reaches only a short distance into the solid shell before the wave has passed. It seems that because the outer core is liquid, the effects that are caused by the wave of pressure in it lag behind the sun and moon much less than the effects previously described in the solid shell which the same forces cause.

The net effects in the outer core are very complex, particularly when variations through the month are considered. *(A special study of these variations should make an interesting project.)* This cycle of frequent compression and relaxation generates heat throughout the outer core in very small amounts and also in the layers of the shell that are affected.

The spacing and amplitude of the waves vary through the month as the relative direction to the sun and moon change. Because the outer core is liquid, rigidity is only a slight factor in the effects within it, so the effect of any new force spreads very rapidly. The effect is cumulative for the full depth of the outer core.

The outer core contains a lot of heat as heat of fusion. It may contain a lot of crystals whose proportion to liquid varies with changes in the rate of supply of heat and the associated changes of temperature. The amount of heat of fusion involved in the temperature changes would serve to smooth out the changes in temperature and would assist the outer core in acting as a giant surge tank to help smooth out variations in the total supply of heat. In the thin layer of the shell that is accessible direct measurements show that temperature increases with depth. The heat must be moving outward by conduction except in the occasional place where liquid is moving.

It seems reasonable to believe that this situation continues throughout the shell. But in the outer core, because it is liquid, convection is probably the main process for moving heat outward. *Would this move the heat outward much faster than conduction?*

Effects in the Inner Core

At 1400 km from the centre of the earth, the earth changes from the liquid state of the outer core to a solid material of much higher density and higher melting range than materials forming the outer core. In this area, there is a considerable increase in the velocity of the P-waves in the upper part of the inner core.

It is suggested that the gravitational attraction of the sun and moon cause heat to be generated in the outer core in two distinct ways:

1. It seems that as the earth rotates, all elements of the inner core must be distorted by the pull of the sun and moon in the same way as in the shell. This process generates heat by internal friction throughout the inner core. The heat moves outward by conduction. The combined pulls of both the sun and moon peaks at new and full moon, when sun, moon and earth are approximately in line. It lags behind the direction to sun and moon by an amount that is probably quite different from the lag in the shell. This is because a great deal of force is generated to displace the inner core, pulling it off centre, and stirring the outer core.

2. The pull of the sun and moon act more strongly on the inner core than on the outer core because it is denser and tends to pull the inner core off centre, as illustrated in Figure 5. Since the outer core is liquid this can occur readily. As the earth rotates, the direction of this pull rotates with it. As a result, the inner core moves within the liquid outer core.

It is suggested that the inner core never gets far off centre because the earth rotates so fast. The movements in the liquid which produce a stirring effect create some heat, through internal friction, in the affected layers of the outer core.

This effect peaks at new moon, when the sun and moon are pulling in the same direction. It is least at full moon, when the sun and moon are pulling the inner core in opposite directions. At other times of the month it varies between these rather wide limits.

Figure 5: Displacement of the inner core.

Cross-section of the earth through the center in the plane of the ecliptic at new moon showing the inner core displaced off center by the pull of the sun and moon.

The displacement is greatly exaggerated.

The direction the inner core is off center is expected to lag by an uncertain amount behind the direction to the sun and moon.

The amount of displacement will vary greatly through the month. Will the angle of lag vary? Will the inner core tend to roll around on the imaginary circle?

If it does, it will probably rotate at a slightly slower rate than the rest of the earth. What will the consequence of this be?

Heat is generated by the stirring in the outer core, caused by movement of the inner core and probably by effects similar to those illustrated in Figure 4 (see page 39).

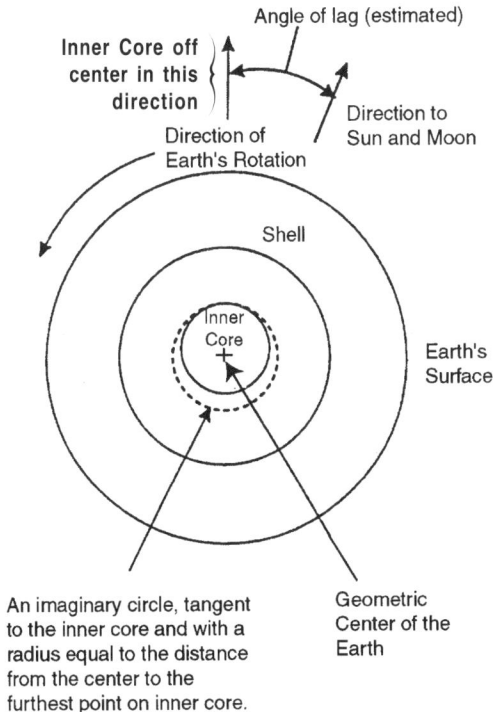

Inner Core off center in this direction

Angle of lag (estimated)

Direction of Earth's Rotation

Direction to Sun and Moon

Shell

Inner Core

Earth's Surface

An imaginary circle, tangent to the inner core and with a radius equal to the distance from the center to the furthest point on inner core.

Geometric Center of the Earth

What the total rate of heat generation is, how it is distributed between these two processes, and how the amount generated by each varies through the month is a matter for more detailed analysis.

The inner core may rotate at a different speed than the earth's surface. This is suggested by examination of Figures 5. The solid inner cores is shown as being slightly off centre. The amount is expected to vary through the month, peaking at new moon and reaching a minimum at full moon. This may result in the inner core rotating at a slightly different rate than the earth's shell — a difference which varies through the month. *It is suggested this is a worthwhile study for someone.*

THE FLOW OF HEAT IN THE EARTH

In this section it is assumed that thermal conductivity is the same all through the area considered.

At places where the heat balance is not exact (i.e., where the supply of heat, and the loss of heat by conduction are not the same), both the temperature and thermal gradient will gradually change to restore balance. This process will start as soon as there is any imbalance and will proceed faster the greater the degree of imbalance.

If there is a surplus of heat the temperature will rise causing the thermal gradient above to gradually steepen, and the rate at which heat is removed by conduction to increase until balance is restored. The increase in temperature will gradually spread upward.

The heat passing through each horizon is diluted going upward because every horizon is larger than the one below. Their areas are in proportion to the squares of their radii. The amount of dilution decreases going upward as represented by the two following examples. Example 1 speaks to what occurs near the base of the shell. Example 2 speaks to what occurs in the shell just below the LV.

Example 1: Near the base of the shell.

All the heat generated below 3499 km from the center of the earth passes through the surface of a sphere of that radius. It continues on and passes through the surface with radius of 3500 km. But the amount of heat it makes per unit of surface area is less because the surface is larger. The areas of the surface vary as the square of their radii. So this heat per unit of area is diluted by 0.057%.

Heat generated in the layer between 3499 and 3500 km in depth goes toward making up this dilution loss. If it is exactly sufficient to make up the dilution, the thermal gradient will be the same at both surfaces.

Example 2: In the shell just below the LV.

Between 5999 and 6000 km from the earth's center the dilution rate is only 0.033 per cent, so the amount of heat that has to be generated in this zone to maintain the thermal gradient is less than in the previous example. Consequently the thermal gradient is apt to steepen as surface is approached.

The important conclusion drawn from Example 2 is that, as surface is approached, heat by internal friction becomes increasingly effective and more likely to pass the IMP.

The following three factors would modify the ideal conditions assumed above:

1. At places where a change of state is occurring, the heat involved in this process substantially affects the heat balance.

2. Thermal gradient is also affected near places where molten rock is moving, since the rock carries with it considerable heat. This is partly due to (a) the high temperature of the rock, and (b) the heat of fusion. Flow in the solid must also have a similar effect but is presumably relatively very slow.

3. Heat received at any spot may not all come from directly below. If the temperature at one horizon varies, heat will move up faster and will tend to spread outward and upward from these hotter spots. This occurrence will tend to equalize the rate at which heat is received throughout the higher horizons and cause an arch in the geotherms above any hot spot. This is particularly significant near areas where magma exists.

The heat appearing at any spot may all be carried away to cooler parts of the shell by conduction. Some heat might go into raising the temperature at the spot, or it might cause melting, in which case an increase in temperature to that area would be slight, because much of the surplus heat would go into heat of fusion.

If melting occurs and the resulting liquid moves, it carries much heat as heat of fusion. Volcanic rocks disperse this heat into the air or disperse it into water after eruption. Plutonic rocks rise to where the cover is thin enough to permit conduction to disperse the heat, and the rocks crystallize.

In other cases, enough heat to maintain temperature might not appear at any spot. In this situation, the spot is cooling and all heat is being carried away through conduction, or is being dispersed by the cooling material. At depth the blanketing effect is high.

The amount of pull of sun and moon goes through a cycle approximately twice a month which causes variations in the rate of generation of heat by internal friction. This period is very short compared to the total time involved. What is to be expected is the effect that the average pull of sun and moon would have. The variations on the effect at depth are only inconsequential on the curve of this effect.

Is a study detailing the amount of the variations throughout the monthly cycle worthwhile?

LAG

If a force is applied to a large mass, there is a lag in time before the mass can complete its response to that force. If the force changes direction during the lag time, the lag changes to suit the new force. A most obvious example of lag is this: The hottest time of the Summer occurs about six weeks after the Summer Solstice. The coldest part of the Winter is about the same time after the Winter Solstice.

This principle of lag applies to the pull of the sun and moon on the inner core. The direction of pull is constantly changing due to the earth's rotation. Further, the direction in which the core is off centre lags behind the direction of the resultant of the sun and moon's pulls as their pull continues to change.

The net lag probably depends on the amount of pull taken up by movement of the inner core which acts as a unit within the liquid outer core, and the amount of pull taken up by deformation of the solid inner core itself.

In the shell this principle applies to the direction of the axes of the imaginary ovoids as noted in Figure 4 *(see page 39)*. They refer to the pull of the sun and moon. The rigidity of the shell causes the axes to lag behind the direction to the sun and moon. *The amount of lag is a matter for further and considerable study.*

In the outer core, lag is expected to be much less than in the shell since material found there is liquid, and the response to the sun and moon's pulls is not delayed to nearly the same extent by rigidity.

The net result is complicated by the fact that the outer core is contained within the solid shell. The pressure wave (resulting from the pull of the sun and moon) which goes around in the outer core will generate a wave going around with it. In this wave material is slightly denser in areas where it is under greater pressure.

Room for a wave of greater volume to go around may be made by a wave of compression in the shell that accompanies the pressure wave. But here too, rigidity comes into play. In the brief time available before the direction of pull changes, it causes a lag in reaction and prevents the compression reaching far into the shell. *What effects do these have on overall results?*

The amount of tilt in the earth due to body tides was measured at one hour intervals during the day and two hour intervals during the night for nearly three months by A.A. Michaelson while doing the research for his 1914 paper. During his studies Michaelson concluded that the pattern of the timing and amplitude of the tides corresponded so exactly with what was expected — due to the changing position of sun and moon — that there was undoubtedly a cause and effect relationship. However, the highs and lows were several hours out of step with what he expected and he could think of no explanation for this lag.

In considering what lag to expect several things must be considered that were not indicated by Michaelson in 1914:

a. The suggestion that the position of the earth's centre of gravity goes through daily movements.

b. The amount of lag differs in the shell, the outer core and the inner core. *The actual amounts in each are at present uncertain.*

c. The actual direction of the plumb line varies instantly, as the centre of gravity moves about within the earth.

d. The apparent direction of the plumb line is affected by the amount and direction of surface tilt, at the moment of observation.

e. The maximum tilt of the earth's surface is along two circles. The circles ring the earth about halfway between the poles and the equatorial plane of the ovoid, into which the earth is distorted by the pulls of the sun and moon at new moon.

If the above considerations are properly allowed for, quite possibly the discrepancies noted by Michaelson, in 1914, will disappear.

EFFECT OF THE EARTH'S ROTATION SLOWING

Flexing and other movements that have been described as creating heat by internal friction throughout the earth are all driven by forces whose reaction tends to reduce the rate of the earth's rotation. Various methods of measuring this have been tried, but results have varied widely and the actual rate of slowing is unknown. However, all methods used to measure rotation seem to indicate the earth's rotation is slowing. Slowing reduces the amount of the earth's kinetic energy.

It is suggested that the amount of kinetic energy lost is the mechanical equivalent of the amount of heat generated within the earth by internal friction.

DISTRIBUTION OF THE SOURCES OF HEAT

Based on the above discussion, the sources of heat within the earth appear to be distributed fairly uniformly horizontally. However, heat's most obvious manifestation on the surface is volcanic action.

A volcano erupts fairly frequently at one place for a period of time and then stops. Later, a period of volcanic action occurs elsewhere.

Periods of mountain building and igneous intrusion are also thought to be surface manifestations of heating at depth, and appear to have a similar distribution in time and space. A hypothesis to account for these discrepancies is offered in the rest of this paper.

THE NEXT QUESTION

The heat is generated within the earth at depth and moves upward by conduction. So, each place above is cooler than the place below.

The history of geologic revolutions indicates that at some places heat raises the rock above the IMP. Any excess heat goes into heat of fusion. But geologic revolutions are periodic events and the ideas presented to this point suggest that heat is supplied pretty uniformly and steadily over time. An explanation of this apparent discrepancy is offered in Part IV.

PART IV: Causes of Subsurface Geological Processes

Summary Outline

Through the years man has observed volcanoes and earthquakes and speculated as to the forces that caused them. Man has also wondered what forces caused intrusive action and diastrophism. To date, no satisfactory theory has been thought of although many have been suggested.

Man has also observed intrusive rocks and wondered how they were emplaced. It is difficult to envision intrusive rocks melting at some spot somewhere at depth and forcing their way up to a point nearer the surface to consolidate. As an alternative, it was suggested that they rose in solution from great depth and replaced the rock that had been where they were. This idea had considerable support at one time.

The recognition of the Low Velocity Zone around 1970 introduced another factor in discussions about what takes place in the earth. The reader will recall, near the end of Part II, the LV was described as a zone in the earth where earthquake waves tend to move slower instead of faster with depth — as happens everywhere else in the shell. Both P- and S-waves pass through the LV, indicating that solid material must be continuous almost everywhere throughout the zone. At all other horizons in the shell, the velocity of both P- and S-waves increases (or at least does not decrease) with depth, indicating that extensive areas that are totally liquid do not exist there.

The hypothesis which follows suggests that heat rising from below causes the temperature in the LV to rise until a liquid fraction forms at some horizons that slows the P- and S-waves. When the temperature rises enough above the initial melting point, some of the liquid fraction becomes continuous. The solid fraction must remain continuous to permit S-waves to pass through.

The difference in density between the liquid and solid results in differences in pressure between the two fractions which forces continuous liquidus up any slope in the zone. It collects at high places into magma which, due to the force of gravity acting on it, rise from there in accordance with the principle illustrated in Figures 1A through 1D *(see pages 18 - 21).*

The behaviour of each magma depends on its density and that of the rocks above and around it. Their differences account for why almost all large intrusives are acidic and most extrusives are intermediate in composition. Also, a mechanism is suggested to explain the forces that cause the movement of material forced in beneath the large areas of marine sediments now found above sea level.

Changes at depth are suggested that account for the starting and stopping.

Details of the hypothesis and what it leads to are set forth on the following pages.

HYPOTHESIS RE THE LOW VELOCITY (LV) ZONE

In developing this hypothesis, it has been assumed that at most places in the LV, most of the earth is made up of roughly horizontal layers of rock. Gravity generally arranges the layers in order of density. The denser layers of rock tend to be more basic, and hence have a higher initial melting point.

Figure 3, *originally found on page 35,* is repeated on the following page, for ease of reference. It shows suggested typical conditions for two things below most places on the earth.

1. **One thing is how the initial melting temperature in the earth varies with depth.** If the rock is the same, the initial melting temperature increases downward. However, it is suggested that the rate of increase is reduced as pressure increases with depth. Above the LV, no attempt is made to show what might happen when going through different rocks, but in the depth range of the LV, steps upward are shown in the curve.

   These steps reflect the assumption that, when going deeper, contacts are crossed, and that a denser more basic rock with a higher initial melting point underlies each contact.

   The steps in Figure 3 are shown to be all the same height and the same distance apart. In fact, these things may vary considerably below any given point, as well as from place to place. In addition, no formation is expected to go all the way around the earth. So, the number of steps varies from place to place.

Figure 3: How Initial Melting Point (IMP) varies with depth.

A generalized picture of how IMP and temperature vary with depth.

Actual position of all these curves and the height of the steps varies somewhat with both time and place.

Estimated temperature and thermal gradients at depth in the earth to the bottom of the Low Velocity Zone (LV), below two places, one with high initial thermal gradient near the surface, the other where it is quite low near the surface.

Figure 3 suggests:

1. *how initial melting point (IMP) varies with depth in the earth.* *The various steps indicate where rock composition changes in the LV.*

2. *how temperature varies with depth in the earth.* *The slope of this line is a measure of the thermal gradient. The curves move up or down with changes of temperature.*

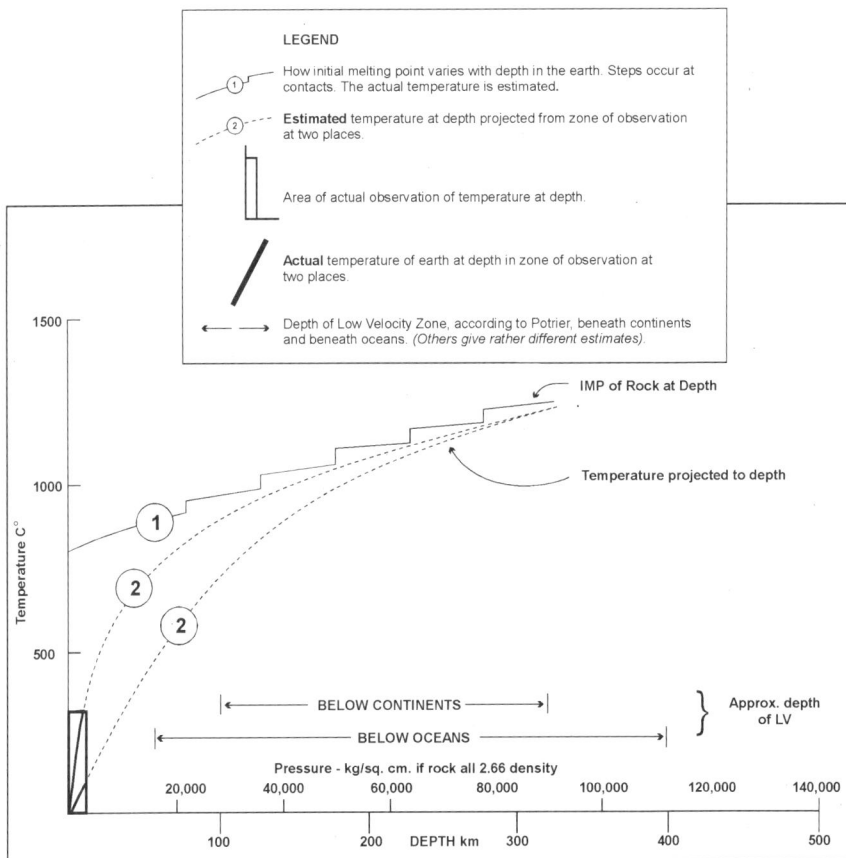

LEGEND

How initial melting point varies with depth in the earth. Steps occur at contacts. The actual temperature is estimated.

Estimated temperature at depth projected from zone of observation at two places.

Area of actual observation of temperature at depth.

Actual temperature of earth at depth in zone of observation at two places.

Depth of Low Velocity Zone, according to Potrier, beneath continents and beneath oceans. *(Others give rather different estimates).*

IMP of Rock at Depth

Temperature projected to depth

BELOW CONTINENTS

BELOW OCEANS

Approx. depth of LV

Pressure - kg/sq. cm. if rock all 2.66 density

20,000   40,000   60,000   80,000   100,000   120,000   140,000

Temperature C°

1500

1000

500

100   200   DEPTH km   300   400   500

2. The second thing presented shows how the temperature of the earth varies with depth. Two curves are drawn to show this. They start from the surface at different slopes. One slope is set by the observed thermal gradient at places where it is quite low. The other is where it is quite high.

If they continued in a straight line, both curves would cut the curve of initial melting point quite soon. Since the earth is known to be practically all solid for about 2900 km, both curves must flatten out enough to do no more than just cut the curve of initial melting temperature. The observed part of these curves is confined to the very small rectangle in the lower left corner of Figure 3 *(see page 35)*. Beyond this, the curves are largely guesswork, subject only to the constraints noted.

Many of our ideas about what the interior of the earth is like come from the study of earthquake waves. As noted earlier, two types of waves, P- and S-waves, travel through the interior of the earth.

The velocity of both waves increases continuously with depth down to the Low Velocity zone (LV), as is expected with higher pressure. In this zone, the increase is interrupted. At most places in the LV, velocities vary with increased depth and pressure, but on the whole, decrease. Beneath the LV, they again increase continuously all the way to the outer core, but the rate of increase is irregular.

According to Potrier *(p.170 et. Seq.)*, the LV zone surrounds the earth. Its depth varies between about 50 and 400 km under oceans, and between 100 and 250 km under continents. Its boundaries are indefinite. At some places the LV seems intermittent. At other places, there is indication of more than one layer. Also, it appears more prominent in areas of magmatic activity. In the old, stable area under the Canadian Shield, the velocity gradient in the zone is nil.

Other writers present a somewhat different picture. However, all seem to agree that the LV exists, that it is very widespread, that P- and S-waves are slowed down passing through it, that its boundaries are indefinite, and that its depth varies from place to place.

As already discussed, it appears that every spot in the crust and mantle below the LV is receiving heat, most of which is generated below by internal friction, and moves upward by conduction. At intervals throughout geologic history, both intrusive and extrusive igneous rocks have formed near or on the surface, from magmas which are believed to have risen from somewhere deeper in the earth. This geologic process confirms that in some places at depth, rock is above the initial melting point. It follows that at these places, heat arrived more rapidly than conduction was able to remove it. The process occurred for so long that the temperature rose above the initial melting point and some liquidus formed.

It is suggested that as the temperature rises, the first liquidus to form will be small spots immediately above a contact where the rock above is less basic, and where it has a lower initial melting point than the rock below. Since heat of fusion now has to be added, the rate of heating here is reduced. This enables the temperature at adjacent places just above the contact to catch up. So the liquidus spreads fairly fast along the contact.

As the temperature rises the liquids will thicken upward away from the contact. The proportion of liquidus will always grade from nil at the top to a larger proportion at the bottom. Everywhere, the spots of liquidus will grow and become more numerous. In time, spots at the bottom will join and the liquidus there will become continuous. Further heating will cause this part to thicken. The proportion of liquidus will always be greatest at the bottom. It will gradually decrease upward to nil at the top.

At places where more heat arrives than conduction is able to remove, the rock gets hotter. This causes the thermal gradient above to steepen, with the effect gradually working through to the surface. So, the rate at which heat is removed increases until a balance is reached.

At some places, heat balance is not reached until after the initial melting point is reached. It also causes the temperature of the rocks below to rise to maintain the thermal gradient in them. So, in total, a lot more heat must be added than is required to raise the temperature at a single spot. But quite a long time will elapse before the upward spreading effect is complete.

After the initial melting point is reached at any place, much of the surplus heat which arrives there goes into heat of fusion. So, the rate at which temperatures rises, drops suddenly. Initially, the resulting liquidus will form small spots scattered through the rock. As the temperature increases, these become more numerous and extensive, and will join to form a continuous network of liquidus in continuous solidus.

Presumably, as the temperature moves further above the initial melting point, the proportion of liquidus increases and the slowing effect on the P- and S-waves increases.

Heat of fusion is an important item in the heat content of magma (*see page 26 — Heat Factors*). In the LV, where pressures are in the order of 25,000 to 100,000 atmospheres, it may form a larger proportion than at surface. *This is discussed låter.*

It seems reasonable to think that when a rock starts to melt, Bowen's Reaction Series for a cooling magma is followed backwards (at least approximately), and that it can be used as a guide to judge the composition of the first fraction to melt. This means that the first liquidus is a good deal more acidic than the average of the surrounding rock. So, its density is lower and its temperature is far below that required to melt the entire rock.

As temperature increases the proportion of liquidus increases, and its composition gradually gets more basic as suggested by following Bowen's reaction series backwards. The density of the liquidus is less than that of the solidus for two reasons:

1. It is more acidic than the solidus.

2. Melting causes some expansion, and therefore reduces density.

Under conditions near the surface, where the rock is acidic, the combined results might amount to a difference of some 15%. The percentage is probably less, the more basic the rock. It may be appreciably different under conditions at the LV, but lacking any guides, it is assumed to be the same.

The presence of liquidus causes the slowing of seismic waves in the Low Velocity Zone. Since S-waves are not transmitted by liquids, but still pass through the LV zone, the solid phase or solidus must be sufficiently continuous and rigid throughout the zone in order to transmit them. It follows that the proportion of liquidus always remains small. Therefore the temperature is never far above the initial melting point.

Melting causes expansion, so each separate spot of liquidus exerts pressure on the material around it.

- Where the liquidus is discontinuous, numerous very small, slow adjustments of pressure, and an increase in volume result all through the rock. The adjustments spread through the rock immediately above and finally reach the surface.

- At areas where the liquidus is continuous, any expansion due to melting is mainly relieved by movement in the liquidus along the LV zone. The liquidus moves to areas where there is least resistance to the upward pushing of overlying rock. Rather than pushing up immediately above every place where liquidus is developing, pools of magma form and expand in areas where there is least resistance.

It is of value to note that seismic evidence shows that the depth to the LV varies. It is further assumed that long, gentle slopes exist in the LV, and that a network of continuous liquidus follows some slopes in the LV.

Before attempting to describe, as a whole, the actual situation in the LV, first consider the theoretical situation that would occur in continuous liquidus existing on a long, uniform, and gentle slope, a slope that is not connected to any other section of the LV and where there is no flow in the liquidus *(see Figure 6A)*.

In this situation, pressure in the solid rock and at every spot in the solidus is equal to the weight of the column of material between it and surface.

Figure 6A: Cross-section of part of an active horizon in the LV.

Cross-section of one horizon in the LV at the peak of a cycle of activity.

Length of area represented is probably at least 5,000 km

and height about 50 km.

Thickness of discontinuous liquidus is pretty uniform as long as there is continuous liquidus below it.

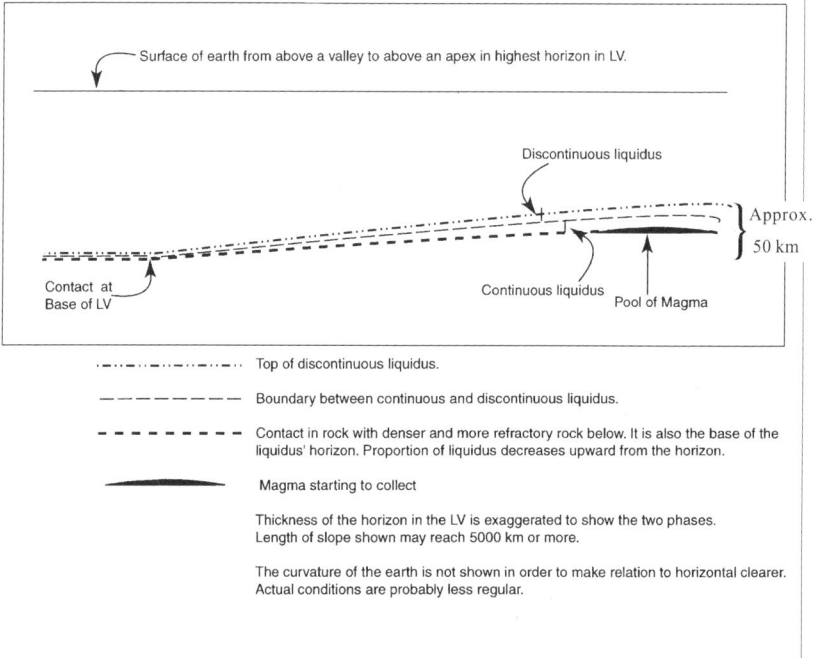

Surface of earth from above a valley to above an apex in highest horizon in LV.

Discontinuous liquidus

Approx. 50 km

Contact at Base of LV

Continuous liquidus

Pool of Magma

Top of discontinuous liquidus.

Boundary between continuous and discontinuous liquidus.

Contact in rock with denser and more refractory rock below. It is also the base of the liquidus' horizon. Proportion of liquidus decreases upward from the horizon.

Magma starting to collect

Thickness of the horizon in the LV is exaggerated to show the two phases. Length of slope shown may reach 5000 km or more.

The curvature of the earth is not shown in order to make relation to horizontal clearer. Actual conditions are probably less regular.

At the lowest point in the LV, pressure should be the same in both phases. If pressure were not the same, either the solidus, the liquidus, or possibly both would slowly move until pressure was equalized. Going up the slope, pressure in the liquidus decreases slower than in the solidus so pressure in the liquidus at any elevation above the lowest is greater than in the solidus and the excess increases going up the slope. *(The principle that has been presented in the previous paragraphs is illustrated by Figures 1A through 1D; first introduced to the reader on page 18 - 21).*

For conditions in natural rock, the graphs presented in Figures 8A, 8B and 8C *(found later in this paper in the section titled: Speculation as to How Magma Pools Expand, page 70)* are more appropriate. This is because the differential between density of liquidus and that of solidus is much less than has been considered in the materials illustrated by Figures 1A through 1D.

In an actual situation, on the single slope the differential pressure will cause the liquidus, wherever it is continuous, to flow up the slope. The continuous liquidus will make room for itself by pushing the roof up where it is easiest to do so. It will eventually accumulate there, forming a pool of magma. Presumably this pool will occur at, or very close to the apex, where pressure in the solidus is inclined to be lowest, and excess pressure in the liquidus is greatest.

The fact that flow is taking place makes the system a dynamic one. Actual pressure in the liquidus is reduced below static pressure by friction losses that result from flow. In Figure 6B, this lies between "static pressure in liquidus" and "pressure in solid rock", but is probably closer to the latter.

Under these conditions, the actual differential pressure is not great anywhere in the LV and may be greatest at some place other than the apex. Indeed its location probably changes as conditions change due to the processes described above, in addition to other processes to be noted later.

Figure 6B: Pressure related to a granitic intrusive.

Relation between pressure and depth in:

1. the elements in a high horizon in the LV from a valley to an apex, and

2. the rocks all the way to surface, and

3. the magma that rises from the apex.

The slope of the lines is set by the density. The rock density is all assumed to be 2.66 and the liquidus and magma to be 2.42.

The depth of the valley is probably somewhat over 100 km, so rock pressure at the valley is somewhere around 25,000 kg/sq cm.

Detail in Figure 6B is conjectural and would vary from place to place, but the broad idea is correct.

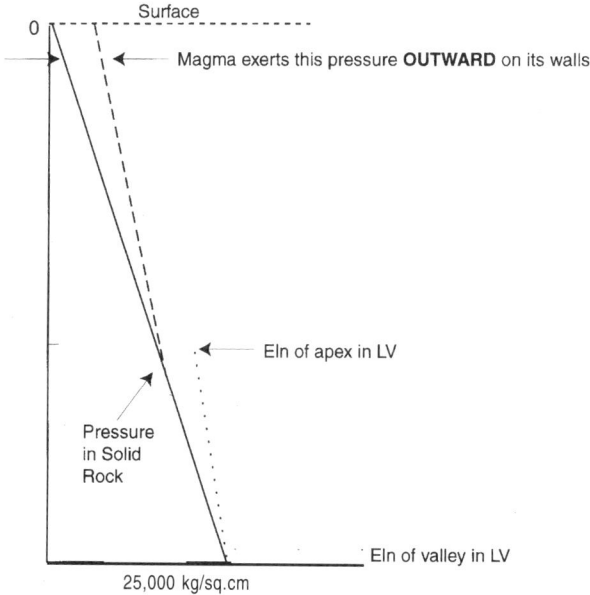

Surface

0

Magma exerts this pressure **OUTWARD** on its walls

Eln of apex in LV

Pressure in Solid Rock

Eln of valley in LV

25,000 kg/sq.cm

Legend:

Pressure in rock. Rock is all assumed to be granite with a density of 2.66.

Pressure in magma. Its slope is the same as that of the imaginary line showing static pressure on the liquidus.

An imaginary line showing what the pressure in the liquidus would be if there were no flow (i.e., the static pressure). The actual or dynamic pressure in the liquidus is less than this because of friction losses incurred in flow and scarcely above rock pressure.

At and near the bottom of the slope in the LV, the supply of heat from below plus what is generated at the spot must exceed the amount lost by conduction. This must occur in order to cause the additional melting needed to supply new liquidus that will replace any liquidus which has moved up the slope.

Failing this, the solidus would close in and reduce the rate of flow, until balance was reached. However, this may not apply in the upper part. There, the amount of heat carried by the "in-flowing" liquidus, much of it as heat of fusion, and the effects that accompany a reduction in pressure, make a difference. Also, the rate heat flows in from below may be less for a number of reasons which will become apparent later.

The proportion of liquidus in the LV is controlled by the rate at which excess heat develops in the area. The excess heat goes partly into raising the temperature of the whole, and partly into supplying heat of fusion for newly melted material. If the rate at which liquidus flows is less than the rate at which heat is generated, the proportion of liquidus increases. An increase in liquidus causes the channels to enlarge. Larger channels reduce frictional resistance and permit the rate of flow to increase. If the rate of flow exceeds the rate of generation, the reverse occurs and thus a balance is maintained.

Liquidus is generated at many places along the slope in the LV, with any excess moving up the slope. At every place along the slope, the rate of flow has to be great enough to remove the excess liquidus generated there, plus what flows to that area from further down the slope. So, the volume of liquidus passing every place where liquidus is being generated increases, going up the slope.

To accommodate an increase in the volume of liquidus, the size of channels increases, or the pressure gradient gets steeper, or a combination of both size and gradient change occur. This would cause the LV to have a greater effect on seismic waves. This means that a horizon containing continuous liquidus normally gets thicker, and/or contains a higher proportion of liquidus, as an apex is approached. However, the liquidus cannot get above some quite modest proportion without the passage of S-waves being stopped.

The size of a channel might be increased by liquidus pushing the walls apart, or by a temperature increase that causes a larger proportion of solidus to melt. An increase to the thickness of the continuous liquidus also requires a temperature increase in the area. Any of the changes described in the preceding paragraphs would occur very slowly.

The picture presented so far has been developed for a single sloping section of LV, as if it existed in isolation *(see Figure 6A on page 64)*. But in actuality, such a section is quite probably connected to many other sections. Sections would be at various slopes, from flat to gentle, in varying directions, and possibly would contain some short, steep sections.

In presenting this picture, another modification to be considered is that heat may arrive too slowly at some places to maintain continuous liquidus everywhere.

Indeed, there may be some areas where discontinuous liquidus is not even found. Consider also that conditions will change with time (see *Changes with Time starting on page 84)*.

To summarize, seismological evidence indicates that the LV is some 200 to 300 km thick, that the top may be less than 100 km below the surface, and that these distances vary from place to place. This suggests that there are several horizons of liquidus. Each is probably above a contact where the rock below is more basic and has a higher initial melting point.

The movement of liquidus in any horizon moves heat from where it leaves to wherever it goes. This reduces the amount of heat available to go upward from lower parts of the horizon. Therefore, the likelihood that continuous liquidus is present in more than one horizon, and beneath any one spot, is also reduced. However, horizons with discontinuous liquidus may exist there, and since they do not remove heat from the area, they affect higher horizons only to the extent that they may absorb heat as heat of fusion if their temperature is rising.

Each horizon yields liquidus and magma of a composition and density that varies with the composition of the zone. Since the more basic rocks are expected to be deeper, the more basic magmas originate in deeper horizons.

This hypothesis suggests that liquidus forms everywhere, or certainly almost everywhere, around the earth in the LV, and that it flows along the LV to certain places where it collects as magma. Magmas appear to occur only at widely spaced intervals, which implies that the area of LV supplying liquidus to any one pool forms an appreciable proportion of the earth. The proportion doubtless varies greatly.

IGNEOUS ROCKS AND DIASTROPHISMS

Speculation as to How Magma Pool Expands

As more liquidus is forced into the pool, it enlarges at a rate that corresponds very closely to the rate of generation in the LV. Conceivably, the pool may take any number of shapes. All of these shapes result in some uplift in the overlying rocks, with consequent changes in pressure in the pool. To get an idea of how the magma is apt to behave, consider the theoretical results if the magma pushed up in shapes which were similar to those shown in Figures 7A through 7C. In each figure, the magma is assumed to have reached the same height. In Figure 7A a certain shape is assumed for the magma.

The displacement it has caused in the overlying rock is assumed to have spread at a uniform angle all the way to the surface. In Figure 7B, the magma is the same shape, but the angle of spread is smaller. In Figure 7C, the width of the base of the magma is half what it is in Figure 7A, but the angle of spread is the same as in 7A.

In these three figures, the solid and dotted lines are presented in pairs. The lowest dotted line represents the top of the zone of continuous liquidus that was in the LV prior to any gathering into a magma chamber. The solid line above it is the top of the magma chamber at the time represented in the figure.

In each pair above the lowest the dotted line is the location of an imaginary horizontal surface in the solid rock, prior to any

Figures 7A - 7C: The shape magma might take as the pool expands and the effect on the rocks above.

The imaginary drawings in Figures 6A - 6B serve to illustrate the shape magma might take as the pool expands.

The figures are theoretical drawings and illustrate the argument that is used in developing the hypothesis about how a pool of magma in the LV is apt to grow at an early stage.

NOTE: An analysis based on these drawings says none of these conditions actually exists.

Fig 7A

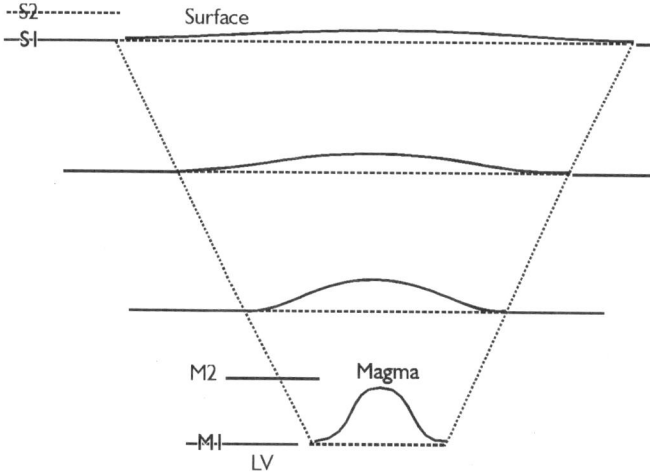

magma forming. The solid line represents where the intrusion moved that surface to. In each pair, the area between the two lines is the same because an equal space has to be provided at each horizon, to make way for the increase in volume below.

As the magma rises, pressure changes take place. In every case, pressure at the base of the magma (M1— the apex in the original

Fig 7B

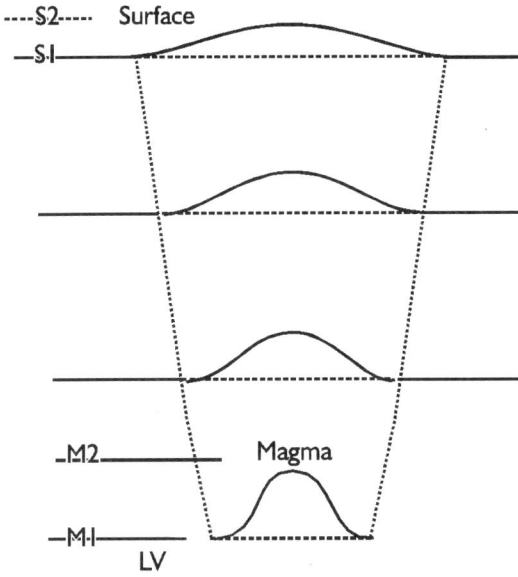

Fig 7C

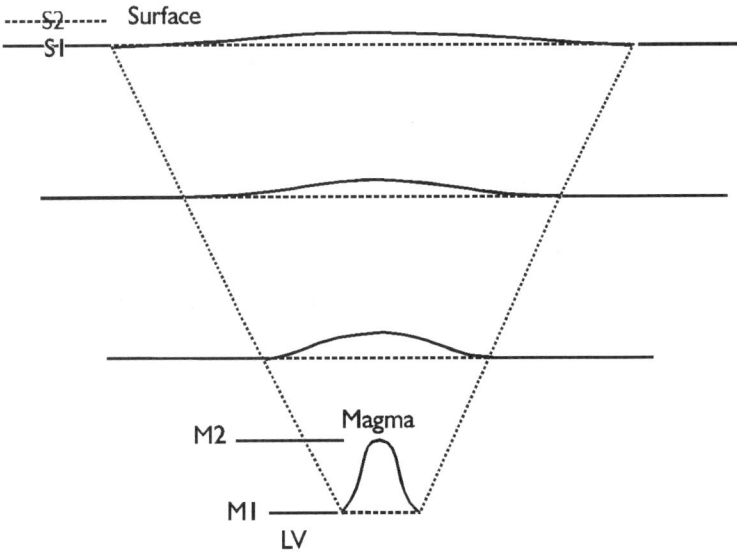

continuous liquidus) is increased by the weight of the column of rock above the original surface. It is reduced by the difference between the weight of the magma and the weight of the rock that it has pushed out of that area . It is estimated that about 85% of the rock's weight has been replaced by the magma.

In Figure 7A there is a moderate net increase in pressure in the LV at the original apex (M1). In Figure 7B, the increase is considerably larger. But in Figure 7C the pressure that existed at the base has been moderately reduced. As soon as the pressure at the base increases, it becomes easier for the liquidus to push the roof up at some other place, where a new pool of magma will develop. In time, this process will be repeated elsewhere.

Clearly, intrusions like those in Figures 7A and 7B would have stopped before reaching the stage shown, and pushing up would have started somewhere else. But an intrusion like 7C has not been stopped yet.

From these results, it is concluded that the wider the angle of spread (Figure 7A vs 7B), the further an intrusion can rise. And the narrower the base (Figure 7C vs 7A), the further it can rise. A dike would rise a long way. A great variety of shapes and sizes of magma, densities of the magma, and overlying rocks could exist. Also the angle of spread may vary on the way up.

The figures presented represent an idealistic situation. However, the following conclusions can be derived with apparent validity.

1. The narrower the base, the faster the top of the magma rises.

2. Magma heats the rocks around it through conduction. The wider the magma, the more chances for pre-heating. The roof and walls of a batholith receive a lot of pre-heating, whereas the walls of a dike receive little. This explains why chilled margins are so much more prominent in dikes.

3. The deeper the active horizon in the LV, the less the rise at surface, and the more extensive is the area that rises. This affects the change in pressure at the active horizon (M1) due to change in the elevation of the surface above. This is illustrated by considering the differences that would result if the surface were at any two of the different elevations in Figures 7A - 7C.

Conclusion as to How Magma Pool will Expand

The actual behaviour of an individual magma is greatly affected by the following variations that were not considered in the ideal case discussed on the previous pages. They are:

1. **The density of the magma.** This is affected by several factors of which the most important is probably:

   ■ its composition. The composition of magma, in turn, depends on the composition of the solidus from which it originated. In general, denser rock, which is also deeper in the earth, is more basic and therefore yields denser magmas.

   ■ its pressure and temperature. At depth, the pressure and temperature of the walls are not very much different from those in the magma, and therefore may not be an important factor.

■ time. As time progresses, crystallization starts and presumably causes an increase in density with the ensuing changes and effects on pressure. Crystallization may not start until quite late in the process of intrusion.

2. **The density of the overlying rock.** The upper part of most places in continents and continental shelves is composed of sediments whose density varies with composition, degree of compaction, proportion of water, and other variables.

   Here, layering or stratification is prominent in the overlying rock. Beneath, there are crystalline rocks of varying density. In these rocks, layering is less prominent among volcanics, and probably almost inconsequential in plutonics of the type that erosion has exposed. Also, only a small fraction of the depth to the base of the magma is represented. The fraction varies and is dependant on that part of the LV from where magma originates.

3. **The elevation of the surface.** The effect of differences in elevation spreads downward. So, at the depth reached by a magma at an early stage, pressure is due to the average elevation of a considerable area, which increases with the magma's depth. As the magma rises, the area which affects pressure on the magma decreases. Near the end of the process, relatively small topographic features may be significant.

4. **The temperature of the rocks being intruded.** Rock temperature increases with depth. It is also higher near other intrusives. The larger and more recent the intrusive, the greater the pre-heating of the host.

5. **Rate of flow.** Rock above a point in the LV, through which liquidus passes, receives some heating from the liquidus. So, the greater the rate of flow the hotter the rock will be, and vice versa.

The five considerations noted on the previous page suggest that a magma pool is most likely to develop and expand as quite a thin layer along the LV, with narrow dikes advancing upward one after another from the LV. The order in which the magma spreads would depend on the way pressures changed with time as the magma spread, the slope of the horizon it is following in the LV, and probably other things that, as yet, have not occurred to the author.

Magma will always spread in the place of least resistance. Changes caused by the spreading magma will change the place where spreading occurred. One at a time, dikes would rise from the pool. The dikes would probably be steep, otherwise there would be extensive caving from the hanging wall. Each dike would rise until loss of heat to the walls started the process of crystallization. Crystallization would continue until rising stopped. Each new dike would preheat the walls a little more, enabling the next dike to rise even higher before crystallization stopped it.

The behaviour of each magma depends mainly on its density. The denser the magma, the deeper the horizon from which it originated in the LV. Typical results for granitic, basaltic, and more basic magmas are suggested in the next few paragraphs.

In the case of a granitic magma, whose density is about 2.42, excess pressure at its top, over pressure in the containing walls, will increase fairly rapidly as the magma rises *(see Figures 8A - 8C, pages 78 and 80)*. But pressure in the surrounding rock decreases at a faster rate. At some point, excess pressure will exceed the pressure in the walls. When this happens, magma, in addition to moving upward, will start to push out sideways.

Since the rate at which magma is supplied remains fairly steady, the rate at which the top rises will slow down. The expanding top will cause heat, through conduction, to dissipate at a faster pace. When this occurs, crystalization begins soon. The process of crystallization slows the intrusion, and eventually stops it. When one intrusion stops, another starts near the apex in the LV.

The situation described continues as long as this horizon is active. If the horizon remains active, other dikes from the same pool will start to rise, and the process repeats itself. The resulting intrusives will often merge into a large batholith. Younger intrusives, in general, rise higher than older ones. As a result, the proportion of intrusives exposed in an area will increase as erosion proceeds. This explains why granitic rocks are so abundant in many parts of the Canadian Shield where erosion is deep.

Near the surface, the pushing apart of rocks on the sides of a batholith causes grabens and structures like the Rocky Mountain Trench to develop directly above the batholith. This also causes mountains like the Rockies to develop on the sides of grabens because of the sideways expansion of the batholith. *Why are mountains like the Rockies not found west of the Trench?* It is suggested that at one time mountains like the Rockies did exist west of the Trench. It is suggested that intrusion was more active west of the Trench, and that batholiths rose higher there. This moved the sediments so high that erosion totally removed them.

Figures 8A - 8C: The relation of pressure and depth in Acidic, Basaltic, and Ultrabasic Magmas.

The following diagrams explain why some magmas form intrusives, some form extrusives and some form sills. It follows the principle illustrated in Figure 1A (see page 18).

The slope of each line indicates the density of the liquid or solid it represents.

Differences between actual densities has been somewhat exaggerated for clarity.

Density of each magma depends on the rock from which it is derived. It is the same all the way to the surface. The denser rocks occur deeper and yield more basic magmas.

Detail in Figures 8A - 8C is conjectural and would vary from place to place, but the broad idea is correct.

Fig. 8A
Acid Magma

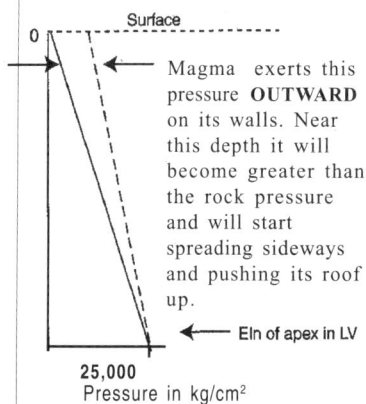

Surface

0

Magma exerts this pressure **OUTWARD** on its walls. Near this depth it will become greater than the rock pressure and will start spreading sideways and pushing its roof up.

← Eln of apex in LV

25,000
Pressure in kg/cm²

Fig. 8B
Basaltic Magma

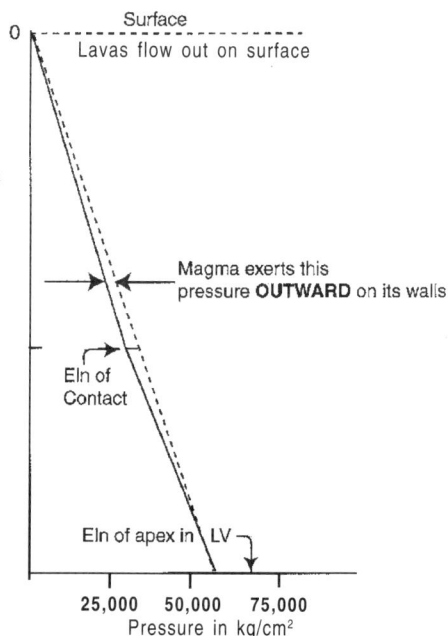

Surface
Lavas flow out on surface

0

Magma exerts this pressure **OUTWARD** on its walls

Eln of Contact

Eln of apex in LV

25,000   50,000   75,000
Pressure in kg/cm²

No reason has yet occurred to the author to explain why intrusion should continue longer at one place than another. *This is a problem for others to consider.* However, igneous activity did continue westward through much of British Columbia. So, the trench is on the edge of the active zone.

At greater depth, on the walls of a granite batholith, where the high pressure makes the rise that accompanies mountain building more difficult, the pressure in the magma squeezes the wall rocks. This pressure is relieved by the wall rocks being squeezed out upward, and causes the steep schisting so common in parts of the Canadian Shield.

If the magma comes from an intermediate horizon in the LV and is basaltic with a density of approximately 2.67 (which is about the same as the upper layers of the acidic crust), it has little outward push *(see Fig. 8B)*. In many environments the magma usually flows out onto the surface as a plateau basalt. Only a small proportion of magmas of approximately this density erupt as volcanoes. This explains why acid rocks are dominant among intrusives and intermediate rocks are so much more common in volcanics.

Basic sills exist at many places. When they were intruded, they must have raised the surface above them. It is suggested that these are from magmas whose density was too high to enable them to reach the surface *(see Fig. 8C)*.

Intrusions of this nature are the cause of large areas of marine sediments, now being found far above sea level. Such a magma is from a very deep horizon in the LV, from a horizon where the rocks

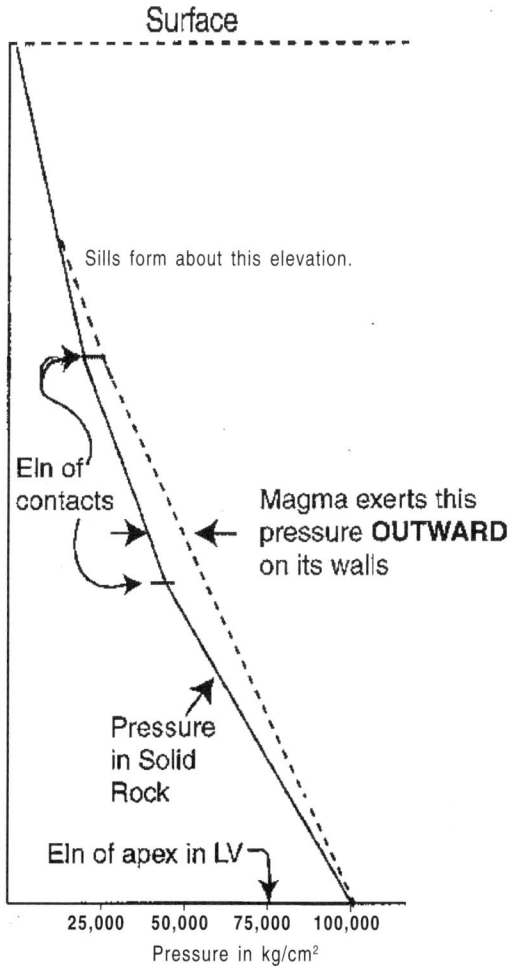

Fig. 8C
Ultrabasic Magmas

Surface

Sills form about this elevation.

Eln of
contacts

Magma exerts this
pressure **OUTWARD**
on its walls

Pressure
in Solid
Rock

Eln of apex in LV

| 25,000 | 50,000 | 75,000 | 100,000 |

Pressure in kg/cm²

are denser and more basic than those in the horizon that yielded the lavas. Sills formed from them are probably, in many cases, so far below the surface that erosion will never expose them.

EFFECT OF PRESSURE ON HEAT AND FUSION

Liquidus is pictured as forming everywhere, or nearly everywhere, around the earth in the LV. It moves upward, along long gentle slopes at various horizons in it, and collects into magmas at or near high places in the LV. The magmas are pictured as rising slowly through progressively cooler rocks. They form relatively small quantities compared to the cooler walls. It seems remarkable that they do not crystallize sooner. The following suggestion may explain why.

It is suggested that the heat of fusion increases with pressure. This is plausible, since the expansion that takes place on melting occurs against progressively higher pressure with depth. So, more work takes place to effect the melting. As the magma rises and pressure decreases, part of the excess heat of fusion is dispersed. Added to heat given off by cooling, this excess heat of fusion heats the walls around the rising magma. As a result, the magma remains molten for a longer period of time.

The additional heat needed to cause fusion deep in the earth is gradually returned to the earth as the liquidus and magma rise into areas of lower pressure. So the total heat required by the processes described is not increased by this concept.

REASONS FOR PRESSURE SURGES

A volcanic eruption occurs at the surface. In this hypothesis, the eruption is pictured as occurring at the top of a column of magma. Probably dike-like in shape, the column is pictured to be over 100 km high extending nearly vertical to a horizon in the LV, a horizon which yields liquidus of intermediate composition.

Liquidus continues downward along the horizon on a gentle slope for a long distance (probably several thousand kilometres), to a low point in the horizon. Liquidus is being generated in the horizon, especially in its lower part. So, the whole horizon is under pressure from its bottom, for reasons explained earlier in the paper.

If the liquidus were completely free to move in response to the additional liquidus being generated in the lower part of the system, there would be a small stream of magma continually emerging on the surface. But it seems that in many cases, something near the surface must offer resistance and cause pressure to build up slowly all through the lower, or bottom part of the system. The increased pressure causes a slight increase of density, and a slight increase in the total mass of liquidus existing throughout the system. It also causes a slight compression of the walls surrounding the liquid, creating more space that can be occupied by the liquid.

When this resistance is finally overcome, expansion occurs suddenly. It starts with a burst of magma erupting from the volcano at the place where initial resistance occurred. This is due to a sudden, rapid expansion of the top of the column of magma, and a slight inward closing of the walls surrounding the liquid in response to the reduced pressure. Next, a wave of expansion moves down through the column. Expansion in the lower part of the column of magma causes the pressure in the upper part of the column to build up again, causing another eruption. As the wave moves downward, this process is repeated, causing a series of eruptions which continue until the wave reaches the bottom of the magma. Since magma is liquid, this process occurs fairly quickly and results in the series of eruptions, usually of decreasing magnitude.

When the wave of expansion reaches the bottom of the magma, it continues along the LV, but at a rate greatly slowed by the dampening effect of very small, crooked channels in the LV. This wave is finally stopped by new liquidus being forced up the slope in the LV. This builds pressure back along the LV, and then in the column of magma, until it rises high enough to cause another series of eruptions. The period of time between one series of eruptions and the next is the period of quiescence of the volcano.

How far down the LV does the wave travel before it is stopped by liquidus under pressure from the bottom? The author speculates that the distance is not great because it is probably several thousand kilometres to the lowest point in the LV.

It was suggested earlier in this section that most magmas that rise from the LV do so along dikes, that these dikes rise higher at certain places along their length, that a volcano occurs where a dike breaks through to the surface, and that at most places the dike may be a considerable distance from the surface. Much volcanic rock exists as plateau lavas, surfacing and flowing over a very large area through fissures which may travel for a great a distance before reaching the surface.

The difference between volcanics and plateau lavas lies only in how extensively the dike they rise through has developed.

Earthquakes usually occur as one major shock, followed at close intervals by a series of aftershocks of decreasing intensity. Then, after a much longer quiet period, there is another big earthquake followed by another series of aftershocks. The similar patterns of time recorded during volcanic eruptions and earthquakes suggests some relationship between the forces that cause them.

It is suggested that earthquakes are caused this way. Magmas are being emplaced at depth to form batholiths. They push their walls aside and roof up, setting up stresses in their walls. In time the stress becomes more than the rock is able to bear. It fractures, causing an earthquake.

CHANGES WITH TIME

Several changes that take place over long periods of time affect the progress of the processes of igneous action and diastrophism. The most important of these processes is the changes that take place in a horizon in the LV where liquidus is forming. Such changes cause the rate of supply to rise to a maximum, then decline, and finally stop.

At one time during this process, there exists only discontinuous liquidus. Heat arrives faster than conduction removes it. Consequently, the temperature rises. In time, a little continuous liquidus appears and immediately starts moving up the slope. However, the rate of movement is very slow because (1) the channels are very small, and (2) frictional resistance is high.

As the temperature continues to rise, the proportion of liquidus increases and the thickness of the zone containing the continuous liquidus increases. Both volume of liquidus and thickness of the zone restrict flow, but become less important with time.

The liquidus is more acidic than the average of the rock, and with time the residual solidus becomes more basic, causing its initial melting temperature to increase. This increases overall temperature, steepens the thermal gradient to the surface, and increases the rate of heat loss through conduction. Gradually, less and less heat is available to generate liquidus.

The process continues at an increasing rate until the amount of heat being carried out by the liquidus equals the amount of heat arriving. At this point in the process, the increasing refractiveness of the remaining solidus means the temperature has to generate more liquidus. Also, the rate of heat lost through conduction increases. So, the rate of generation of liquidus starts to slow. Since the rate of the liquidus' outflow tends to stay in balance with the rate of its generation, the process finally reaches a point where there is no continuous liquidus, and flow stops.

When flow stops, conduction has removed all the heat arriving and balance is attained. Only discontinuous liquidus remains. Nothing changes until something (another geologic process) occurs to disturb the balance.

The processes described on the previous page develop at a low point in the LV if, as explained, an entire cycle is completed. This fundamental cycle, which every low point in the LV goes through, is important because it accounts for why geologic revolutions start and stop.

In practice, changes in temperature above and below the LV may result in another horizon becoming active and interfering with the cycle in the first horizon. Going up a slope in the horizon, the rate at which higher temperature liquidus is supplied from a lower point on the slope complicates developments. Complications vary with the rate of flow, but the new horizon must go through the same cycle as the first horizon. Similar cycles may be occurring at other places in other horizons, any of which may interfere with the smoothness of the fundamental cycle in the first horizon, and vice versa. The low points are probably not one above another, but are widely scattered.

To re-emphasize, this fundamental cycle which every low point in the LV goes through is important because it accounts for why geologic revolutions start and stop.

CYCLE IN CHANGES WITH TIME

The fundamental cycle described above affects the rock above and below a horizon. At both places, rock gets hotter as the cycle progresses, and liquidus may start forming a new horizon. Let us call the first, or original horizon, "B". The horizon below "B" is called "C", and the horizon above it is called "A".

If there is a low in horizon "C" (which is directly below the original low), and liquidus starts to form there, its formation will take up heat of fusion. This process will reduce the amount of heat available to move the liquidus up from "C" and will interfere with the cycle in "B". At a certain temperature, the liquidus becomes continuous and starts to move. So, the fundamental cycle which started in horizon "C" will carry heat out of the area. As the temperature in horizon "C" increases, it interferes with conditions in horizon "B", causing "B" to peak sooner and at a lower temperature. If "B" is past its peak the cycle will end relatively abruptly.

If there is a low in horizon "A" (directly above the low in "B"), the temperature at "B" rises such that the thermal gradient from it to the surface will steepen. Thus the temperature at "A" will rise, but not as much as the temperature at horizon "B". The closer the horizons, the less the temperature differs.

Of course, there will also be some lag to consider. But because the cycle is slow, the lag may not be large. Over a period of time, this horizon may enter the same cycle. The sequence and timing of when "A" and "C" enter the same cycle varies with the composition of each horizon, the distances between the horizons, and possibly other considerations not mentioned here.

It is much more likely that the lows in horizons "A" and "C" will not be directly above or below those in horizon "B". They will instead be below some other point.

The processes described in the fundamental cycle can be easily modified to show what happens at site conditions, where the lows and highs are not directly below one another. Also, changes that are probably quite small and which occur at places where activity began during a period when a horizon was active, may occur and have effects.

The entire operation is a very slow process, because all the rock — from the surface down to at least the lowest horizon — must be heated, not just the rock in the active horizons.

THE VOLUMES THAT MOVE

My knowledge of historical geology is slight — just what carries over from my student days in the thirties. Based on this the following suggestion is offered.

A large part of central North America is underlain by flat-lying sediments — mostly marine. Study of these indicates that parts of the area have fluctuated from being above and below sea level. From this it appears that sometimes material has been forced into the area as deep sills. At other times material has been taken out — presumably as liquidus in the LV. A casual estimate of the volume now above sea level made some time ago suggests it is over one million cubic kilometres. Also the distances most of it move must be at least many hundred kilometres to enable it to get in beneath this large area from outside it or vice versa.

Since there appear to have been no more than slight flexing, the entry and withdrawal must have been spread pretty uniformly over large areas.

The large volumes and distances involved make it likely that material that moved was in the liquid state.

It is suggested that the material withdrawn was liquidus moving in the LV and the material entering was in sills.

EFFECTS OF CHANGES ON THE SURFACE

The pressure at any point at depth decreases when material from the surface above that point is removed. It increases with the deposition of material there. The temperature in the area is also affected. If material is removed, the temperature at the surface rises because heat generated from below passes through less rock. If material is deposited, blanketing at the surface increases, and the temperature below the surface rises.

The affected area at depth will spread downward, and both the size of the area and the heating effect in the area will progressively be reduced with depth. So, the effect at any place in the LV will only be significant, if the area affected on the surface is large.

At the surface the effect of erosion and depositing will be equal to the weight of the material removed. But the effect of submarine sedimentation will equal the weight of material deposited, less the weight of water displaced. Calcitic material deposited in water adds its weight, less the weight of water it displaces, to the load on the area below it. But, it comes from material in solution in the water, so its lightening effect on the load under any particular land area is nebulous.

PART V: Conclusion

The processes which occur in the Low Velocity Zone and the heat sources within the earth offer an explanation for geologic revolutions and processes that present day thinking has failed to explain. The hypothesis presented in this paper differs from conventional thinking.

Summary of Modifications to Conventional Thinking

Significant modifications or additions to conventional thinking have been proposed in this paper and are summarized below.

- Radioactive decay is reduced by pressure and is nil below a few tens of kilometres of depth in the earth.

- The proportion of radioactive material near the surface of the earth is maintained by new material being brought up by magmas from depth.

- In order for life to have developed as it has, water must have existed on the earth's surface for a very long period of time. So, the earth's surface temperature must have stayed within quite narrow limits during all that time.

- Gravitational attraction of sun and moon cause tides in the solid earth that generate heat by internal friction all through the earth.

This occurrence causes the temperature everywhere in the interior of the earth to rise until the thermal gradient is steep enough to maintain balance at that place.

- The Low Velocity Zone surrounds the earth. It is interpreted as being that part of the earth where thermal balance is reached only after initial melting point is reached — a least at some horizons in the zone. When this occurs, some spots of liquidus appear immediately above a contact. As the temperature continues to increase, the liquidus becomes continuous and gravity forces it up any slope.

- The presence of liquid slows seismic waves. S-waves always pass through the Low Velocity Zone — an indication that the solidus is always continuous.

- Gravity is the force that drives most geologic processes. It has arranged the solid parts more or less in order of density. Solid material does not move easily. But when the temperature at some places rises enough that a liquid fraction forms, the density is reduced. Gravity makes the liquid tend to rise. And its liquid nature enables it to pass through very small openings and to change shape readily to fit the changing shape of the openings.

- When liquidus vacates an area, it leaves behind a solidus that gets more refractory with time causing the process in one horizon to gradually stop. As this cycle occurs, the temperature in horizons above and below the active horizon rises and, in time, one or more of these horizons will become active and the cycle will start again.

- The liquid forced up a slope in the Low Velocity Zone accumulates as a pool of magma at any apex in the horizon. The liquid rises through a dike and its behaviour depends on the density of both the dike and the surrounding rocks.

  - Low density dikes tend to spread out sideways a few kilometres below the surface of the earth, forming batholiths which push the walls of the dike apart.

  - Denser magmas tend to erupt.

  - Magmas which are very dense tend to form sills at depth, raising the surface of the earth above the area of activity.

- The surface area over any place from where liquidus or magma has moved must settle.

- Active volcanos tend to have brief periods of frequent eruptions separated by periods of dormancy. Eventually they become inactive.

- It is suggested that pressure increases heat of fusion. As liquidus and magmas rise and pressure decreases, this extra heat of fusion is gradually released. This idea makes it easier for dikes to stay in a liquid state while rising through cooler rocks.

ADVANTAGES PRESENTED BY THE HYPOTHESIS

The hypothesis discussed in this paper presents the following advantages:

1. Heat is generated throughout the earth.

2. Gravity is the driving force.

3. Pressure at every spot is pictured as being pretty well in balance until a fraction of the solid turns into liquid. This creates a lighter fraction which tends to rise.

4. Liquid is developed all around the earth.

5. A mechanism by which the liquid fraction is collected into a relatively small area and which affects the surface above that area of activity is described.

6. It explains why intrusives are mainly acidic and extrusives mainly more basic. It also explains (but rather sketchily) why magmas that are more basic spread out as sills beneath the surface causing uplift of the central plains of North America and similar areas.

7. It explains why geologic revolutions start and stop at one place.

8. Most of the movement takes place in the liquid state.

The processes described in this paper would seem to produce rather obvious results. In fact the infinite number of complexities to be expected in the earth will produce numerous variations from the ideal picture painted, and account for the wide variations observed on the surface.

There will always, of course, be unanswered questions. It is my sincere hope that the ideas presented in this paper and the questions which arise from them will be studied and further developed.

Acknowledgements

The large amount of typing done by Mr. R.C. Weston and his patience in dealing with numerous modifications, large and small, over a number of years is gratefully acknowledged.

The valuable criticisms offered by Mr. I.S. Parrish have added a lot to the orderliness and clarity of this paper. Discussions with, and information from Dr. William Petrie are appreciated and have aided the author. Also, Mr. Nelson Hogg's criticism has been valuable.

Finally, special thanks to Mrs. Carmel Thomson, of TCL Thomson Communications Limited, who has put a lot of very effective work in the last year into presenting the ideas in a clearer and more logical manner.

References

- BOWEN, N.L., 1920, *Bowen's Reaction Series*, (Experiments at the Carnegie Geophysical Institute resulting in the development of Bowen's Reaction Series).

- BOWEN, N.L., 1928, *The Evolution of the Igneous Rocks*, Princeton University Press.

- GROUT, F.F, 1932, *Petrology and Petrography*, McGraw-Hill, p. 155.

- MICHAELSON, A.A., *1914, Astrophysical Journal*, 1914, p.105-138.

- POTRIER, 1914, A paper written by this Belgian was read by the author. It seemed to contain the best description of the Low Velocity Zone. Unfortunately my notes are mislaid. I think it was a one of a group of papers in one book.

- WYLLIE, P.J., 1976, *The Way the Earth Works: An Introduction to the New Global Geology and its Revolutionary Development*, John Wiley & Sons, p.92 et seq.